Methods in Molecular Biology

Series Editor
John M. Walker
School of Life and Medical Sciences
University of Hertfordshire
Hatfield, Hertfordshire, UK

For further volumes:
http://www.springer.com/series/7651

For over 35 years, biological scientists have come to rely on the research protocols and methodologies in the critically acclaimed *Methods in Molecular Biology* series. The series was the first to introduce the step-by-step protocols approach that has become the standard in all biomedical protocol publishing. Each protocol is provided in readily-reproducible step-by step fashion, opening with an introductory overview, a list of the materials and reagents needed to complete the experiment, and followed by a detailed procedure that is supported with a helpful notes section offering tips and tricks of the trade as well as troubleshooting advice. These hallmark features were introduced by series editor Dr. John Walker and constitute the key ingredient in each and every volume of the *Methods in Molecular Biology* series. Tested and trusted, comprehensive and reliable, all protocols from the series are indexed in PubMed.

Plant-Pathogen Interactions

Edited by

Nora A. Foroud and Jonathan A. D. Neilson

Lethbridge Research and Development Centre, Agriculture and Agri-Food Canada, Lethbridge, AB, Canada

Humana Press

Editors
Nora A. Foroud
Lethbridge Research and
Development Centre
Agriculture and Agri-Food Canada
Lethbridge, AB, Canada

Jonathan A. D. Neilson
Lethbridge Research and
Development Centre
Agriculture and Agri-Food Canada
Lethbridge, AB, Canada

ISSN 1064-3745 ISSN 1940-6029 (electronic)
Methods in Molecular Biology
ISBN 978-1-0716-3158-4 ISBN 978-1-0716-3159-1 (eBook)
https://doi.org/10.1007/978-1-0716-3159-1

Cover Illustration Caption: Fusarium and Aphanomyces root rot of pea samples provided by Dr. Syama Chatterton; photo credit to Dr. Nora A. Foroud.

This Humana imprint is published by the registered company Springer Science+Business Media, LLC, part of Springer Nature.
The registered company address is: 1 New York Plaza, New York, NY 10004, U.S.A.

Preface

This edition of *Plant-Pathogen Interactions* is focused on crop diseases, with the first four chapters dedicated to molecular methods of phytopathogen detection. A protocol and a case study of molecular identification of two of the most important genera of fungal pathogens in crops, *Fusarium* and *Puccinia*, are followed by a chapter on a DNA barcoding method to identify fungal and oomycete phytopathogens and a LAMP assay protocol for in-field detection of the bacterial species *Xylella fastidiosa*.

Cutting-edge methodologies for studying host-pathogen interactions at the earliest stages of colonization is presented in wheat and tomato plants, followed by a protocol to study effector proteins in barley leaves, as well as a perspective on bioinformatics approaches in effector identification.

Following a review on RNA-seq analysis in cereals, two case studies offer unique methods of RNA-seq analyses designed to facilitate interpretation of large datasets when working with complex crop genomes or for improved simultaneous investigations of host and pathogen gene expression. Detailed methods for host responses downstream of, or coincidental to, gene expression (proteomics, fatty acid, and metabolomics analyses) are presented in the next four chapters.

The book is concluded with a newly established protocol for gene editing an important forage crop, the allotetraploid species *Medicago sativa*, where genes and gene products identified using other methods presented in this book can be investigated for their role in crop disease.

Lethbridge, AB, Canada

Nora A. Foroud
Jonathan A. D. Neilson

Contents

Contributors

REEM ABOUKHADDOUR • *Lethbridge Research and Development Centre, Agriculture and Agri-Food Canada, Lethbridge, AB, Canada*
ERIC AMUNDSEN • *Lethbridge Research and Development Centre, Agriculture and Agri-Food Canada, Lethbridge, AB, Canada*
GUUS BAKKEREN • *Summerland Research and Development Centre, Agriculture and Agri-Food Canada, Summerland, BC, Canada*
BRIANNA BALL • *Department of Molecular and Cellular Biology, University of Guelph, Guelph, ON, Canada*
CHRISTOPHER BLACKMAN • *Department of Cell and Systems Biology, University of Toronto, Toronto, ON, Canada*
REID BUCHANAN • *Department of Molecular and Cellular Biology, University of Guelph, Guelph, ON, Canada*
KIMBERLEY BURTON HUGHES • *Lethbridge Research and Development Centre, Agriculture and Agri-Food Canada, Lethbridge, AB, Canada*
ANA RITA CAVACO • *Plant Biology Department, Faculty of Sciences, BioISI – Biosystems & Integrative Sciences Institute (BioISI), University of Lisbon, Lisbon, Portugal*
SYAMA CHATTERTON • *Lethbridge Research and Development Centre, Agriculture and Agri-Food Canada, Lethbridge, AB, Canada*
GUANQUN CHEN • *Department of Agricultural, Food and Nutritional Science, University of Alberta, Edmonton, AB, Canada*
ANTONIO DI PIETRO • *Department of Genetics, University of Córdoba, Córdoba, Spain*
NICOLAS FEAU • *Canadian Forest Service, Pacific Forestry Centre, Natural Resources Canada, Victoria, BC, Canada*
ANDREIA FIGUEIREDO • *Plant Biology Department, Faculty of Sciences, BioISI – Biosystems & Integrative Sciences Institute (BioISI), University of Lisbon, Lisbon, Portugal*
NORA A. FOROUD • *Lethbridge Research and Development Centre, Agriculture and Agri-Food Canada, Lethbridge, AB, Canada*
MICHELE FRICK • *Lethbridge Research and Development Centre, Agriculture and Agri-Food Canada, Lethbridge, AB, Canada*
JENNIFER GEDDES-MCALISTER • *Department of Molecular and Cellular Biology, University of Guelph, Guelph, ON, Canada*
MOHAMED HAFEZ • *Lethbridge Research and Development Centre, Agriculture and Agri-Food Canada, Lethbridge, AB, Canada*
RICHARD C. HAMELIN • *Forest and Conservation Sciences, University of British Columbia, Vancouver, BC, Canada*
ABDELALI HANNOUFA • *London Research and Development Centre, Agriculture and Agri-Food Canada, London, ON, Canada*
MARIA ANTONIA HENRIQUEZ • *Morden Research and Development Centre, Agriculture and Agri-Food Canada, Morden, MB, Canada*
PADMINI HERATH • *Forest and Conservation Sciences, University of British Columbia, Vancouver, BC, Canada*
DANISHA JOHAL • *Department of Molecular and Cellular Biology, University of Guelph, Guelph, ON, Canada*

NANCY P. KELLER • *Department of Medical Microbiology and Immunology, University of Wisconsin-Madison, Madison, WI, USA; Department of Pathology, University of Wiconsin-Madison, Madison, WI, USA*

JESSE LABBÉ • *Biosciences Division, Oak Ridge National Laboratory, Oak Ridge, TN, USA; Invaio Sciences, Cambridge, MA, USA*

ANDRÉ LAROCHE • *Lethbridge Research and Development Centre, Agriculture and Agri-Food Canada, Lethbridge, AB, Canada*

GONÇALO LAUREANO • *Plant Biology Department, Faculty of Sciences, BioISI – Biosystems & Integrative Sciences Institute (BioISI), University of Lisbon, Lisbon, Portugal*

YIFENG LI • *Department of Computer Science, Brock University, St. Catharines, ON, Canada*

BOYAN LIU • *Department of Molecular and Cellular Biology, University of Guelph, Guelph, ON, Canada*

ZIYING LIU • *Digital Technologies Research Centre, National Research Council Canada, Ottawa, ON, Canada*

MICHELE C. LOEWEN • *Aquatic and Crop Resources Development Research Centre, National Research Council of Canada, Ottawa, ON, Canada*

NICOLA LUCHI • *National Research Council, Institute for Sustainable Plant Protection (CNR-IPSP), Florence, Italy*

ANA RITA MATOS • *Plant Biology Department, Faculty of Sciences, BioISI – Biosystems & Integrative Sciences Institute (BioISI), University of Lisbon, Lisbon, Portugal*

DUCCIO MIGLIORINI • *National Research Council, Institute for Sustainable Plant Protection (CNR-IPSP), Florence, Italy*

ANA PRISCILLA MONTENEGRO ALONSO • *Department of Botany, University of British Columbia, Vancouver, BC, Canada; Saskatoon Research and Development Centre, Agriculture and Agri-Food Canada, Saskatoon, SK, Canada*

KIRBY T. NILSEN • *Brandon Research and Development Centre, Agriculture and Agri-Food Canada, Brandon, MB, Canada*

THÉRÈSE OUELLET • *Ottawa Research and Development Centre, Agriculture and Agri-Food Canada, Ottawa, ON, Canada*

YOULIAN PAN • *Digital Technologies Research Centre, National Research Council Canada, Ottawa, ON, Canada*

FRANCESCO PECORI • *National Research Council, Institute for Sustainable Plant Protection (CNR-IPSP), Florence, Italy*

AMEY REDKAR • *Department of Genetics, University of Córdoba, Córdoba, Spain; National Centre for Biological Sciences, Tata Institute of Fundamental Research, Bengaluru, India*

BRONWYN E. ROWLAND • *Department of Biomedical and Molecular Sciences, Queen's University, Kingston, ON, Canada*

ALBERTO SANTINI • *National Research Council, Institute for Sustainable Plant Protection (CNR-IPSP), Florence, Italy*

MITRA SERAJAZARI • *Plant Agriculture, University of Guelph, Guelph, ON, Canada; Ontario Agriculture College, University of Guelph, Guelph, ON, Canada*

TANYA SHARMA • *Department of Chemistry and Biomolecular Sciences, University of Ottawa, Ottawa, ON, Canada*

STACY D. SINGER • *Lethbridge Research and Development Centre, Agriculture and Agri-Food Canada, Lethbridge, AB, Canada*

POOJA S. SRIDHAR • *Department of Biomedical and Molecular Sciences, Queen's University, Kingston, ON, Canada*

UDAYA SUBEDI • *Lethbridge Research and Development Centre, Agriculture and Agri-Food Canada, Lethbridge, AB, Canada; Department of Agricultural, Food and Nutritional Science, University of Alberta, Edmonton, AB, Canada*

RAJAGOPAL SUBRAMANIAM • *Ottawa Research and Development Centre, Agriculture and Agri-Food Canada, Ottawa, ON, Canada*

ANURADHA SURENDRA • *Digital Technologies Research Centre, National Research Council Canada, Ottawa, ON, Canada*

JOANNA TANNOUS • *Biosciences Division, Oak Ridge National Laboratory, Oak Ridge, TN, USA*

MELISSA TELFER • *Lethbridge Research and Development Centre, Agriculture and Agri-Food Canada, Lethbridge, AB, Canada*

DAVID TURRÀ • *Department of Agricultural Sciences, University of Naples Federico II, Portici, Italy; Center for Studies on Bioinspired Agro-Enviromental Technology, Università di Napoli Federico II, Portici, Italy*

SEAN WALKOWIAK • *Grain Research Laboratory, Canadian Grain Commission, Winnipeg, MB, Canada*

Chapter 1

Specific Detection and Quantification of Major *Fusarium* spp. Associated with Cereal and Pulse Crops

Mohamed Hafez, Melissa Telfer, Syama Chatterton, and Reem Aboukhaddour

Abstract

Plant pathogenic *Fusarium* spp. are widespread and cause important diseases on a wide host range, including economically important cereal and pulse crops. A number of molecular methods have been used to detect, identify, and quantify a long list of plant pathogenic *Fusarium* spp. In general, these methods are much faster, highly specific, more sensitive, and more accurate than culture-based methods and can be performed and interpreted by personnel with no specialized taxonomical expertise. The accurate isolation and identification of these pathogens is required to effectively manage diseases caused by pathogenic *Fusarium* spp. In this chapter, we present detailed molecular methods for detection, quantification, and differentiation between many of the *Fusarium* spp. associated with cereal and pulse crops.

Key words *Fusarium*, Molecular identification, Molecular marker, PCR, qPCR, RFLP

1 Introduction

The genus *Fusarium* poses a multifaceted threat to global crop production and animal/human health [1]. Fusarium root rot (FRR) is a major disease of pulse crops and can be caused by a number of *Fusarium* spp., with *F. avenaceum* predominant in recent years, while *F. solani*, *F. redolens*, and *F. oxysporum* are also common [2–4]. On pea and lentil, *Fusarium* spp. often occur as a complex with other soilborne pathogens, such as *Aphanomyces euteiches*, *Pythium* spp., and *Rhizoctonia solani*, making understanding their role in causing primary infection and disease difficult to decipher. Pathogenicity testing of *Fusarium* spp. on pea, lentil, and chickpea indicated that *F. avenaceum* is highly aggressive on all of these hosts, whereas *F. solani*, *F. redolens*, and *F. oxysporum* have a diverse response depending on pulse species [5].

Nora A. Foroud and Jonathan A. D. Neilson (eds.), *Plant-Pathogen Interactions*, Methods in Molecular Biology, vol. 2659, https://doi.org/10.1007/978-1-0716-3159-1_1,

F. avenaceum is also highly aggressive on cereal seedlings [6] and has been associated with *Fusarium* head blight (FHB) on cereals [7–9]. FHB is the most damaging disease of wheat but can also affect other cereal crops like barley, oats, rice, corn, triticale, and rye [10]. FHB is caused by a consortium of mutually interacting *Fusarium* spp., and most FHB pathogens are members of the *F. graminearum* species complex (FGSC). In addition, other species like *F. avenaceum*, *F. cerealis*, *F. culmorum*, *F. poae*, and *F. sporotrichioides* are also associated with FHB [11]. Fusarium crown and root rot (FCRR) is another major disease of wheat worldwide, with *F. culmorum* and *F. pseudograminearum* as the most common causal agents [12]. Several *Fusarium* spp. can combine different lifestyles (saprophytic and/or pathogenic), which makes the interaction between plants and their associated *Fusarium* spp. more complex [13]. Many of the *Fusarium* spp. recovered from diseased plant tissue may be acting as "saprophytes" and are not "pathogenic" [14]. *Fusarium* spp. commonly associated with economically important diseases that affect cereal and pulse crops are indicated in Fig. 1.

Limitations of traditional culture-based, morphological approaches to identify highly similar *Fusarium* spp. have led to the development of molecular approaches with improved accuracy and reliability. Several genes and DNA regions have been used to define species boundaries within the genus *Fusarium* such as the internal transcribed spacer (ITS) region [15], mating type (MAT) locus genes [16], β-tubulin (*tub1*) gene [17], and genes encoding for portions of the DNA-directed RNA polymerase II largest (RPB1) and second largest (RPB2) subunits [18]. However, the translation elongation factor 1 alpha (*TEF1α*) gene has been the most frequently used and reliable sequence to define species boundaries within the genus *Fusarium* [19–22]. Several techniques are used for molecular identification of *Fusarium* spp., including conventional PCR [23], real-time quantitative PCR (qPCR [24]), and restriction fragment length polymorphism (RFLP [25]).

2 Materials

2.1 Media

For all recipes, media are sterilized by autoclave and allowed to cool to ~50 °C before adding antibiotic stock solutions where applicable (*see* **Note 1**). For media containing agar, autoclaved media are dispensed into Petri dishes after addition of antibiotics: 20–25 mL media per 9 cm Petri dish, or 10–12 mL media per 6 cm Petri dish. Large Petri plates (9 cm) are recommended when isolating *Fusarium* spp. from plant tissues to avoid colony overlap, while small Petri plates (6 cm) are suitable for growing pure/single *Fusarium* isolate.

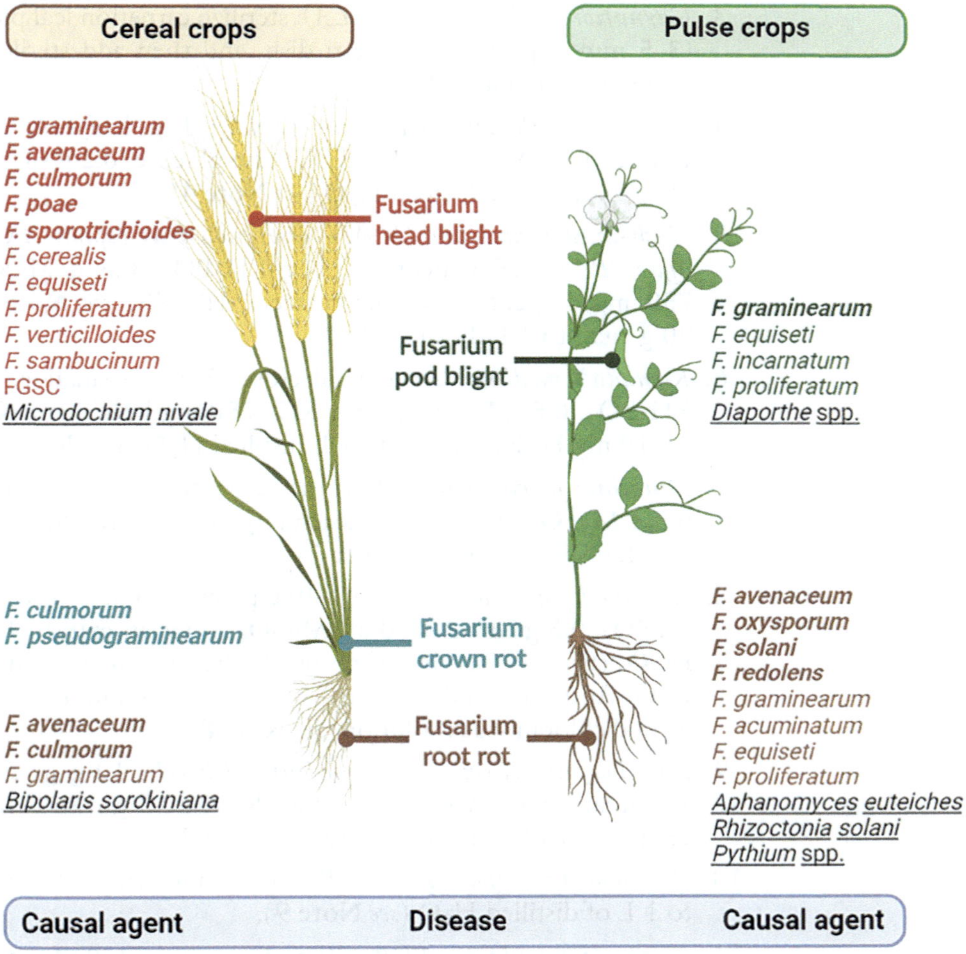

Fig. 1 Many *Fusarium* spp. can infect cereal and pulse crops causing economically important diseases such as *Fusarium* head blight, *Fusarium* crown rot (cereals), *Fusarium* pod blight (pulses), and *Fusarium* root rot (cereal and pulse crops). Commonly associated *Fusarium* spp. with each disease are shown in **bold**, and less common *Fusarium* spp. are also indicated. In addition to *Fusarium* spp., other fungal phytopathogens are also commonly associated with these diseases (*underlined*). FGSC = *Fusarium graminearum* species complex (contains several closely related *Fusarium* spp. that are the primary etiological agent of *Fusarium* head blight of wheat, barley, oats, and other small cereal grain crops worldwide). (Figure generated by BioRender (https:/ biorender.com/))

1. *Antibiotic stock solutions*: streptomycin stock is 5 g of streptomycin in 100 mL distilled H_2O and used at the rate of 20 mL/L of media, and the neomycin stock is 1 g of neomycin sulfate in 100 mL distilled H_2O and used at the rate of 12 mL/L of media (*see* **Note 1**).
2. *Water agar (WA) medium*: dissolve 20 g agar in 1 L distilled H_2O (*see* **Note 2**).

3. *Carnation leaf-piece agar (CLA)*: sterilize carnation leaf pieces (3–5 mm^2), place into a Petri dish, and then add sterile 2% water agar (*see* **Note 3**).
4. *Spezieller Nährstoffarmer agar (SNA)*: 1 g KH_2PO_4, 1 g KNO_3, 0.5 g $MgSO_4$, 0.5 g KCl, 0.2 g glucose, 0.2 g sucrose, 20 g agar, to 1 L distilled H_2O (*see* **Note 4**).
5. *Nash–Snyder medium (NSM), also known as peptone PCNB agar (PPA)*: 15 g peptone, 1 g KH_2PO_4, 0.5 g $MgSO_4$, 750 mg/L pentachloronitrobenzene (PCNB; *see* **Note 5**), 20 g agar, to 1 L distilled H_2O.
6. *Komada's medium*: 20 g D-galactose, 2 g L-asparagine, 1 g KH_2PO_4, 0.5 g $MgSO_4$, 0.5 g KCl, 750 mg PNCB (*see* **Note 5**), 10 mg Fe_3Na EDTA, to 1 L distilled H_2O (*see* **Note 6**).
7. *Malachite green agar (MGA)*: 15 g peptone, 1 g KH_2PO_4, 0.5 g $MgSO_4$, 2.5 mg/L malachite green oxalate, 20 g agar, to 1 L distilled H_2O (*see* **Note 7**).
8. *Selective Fusarium agar (SFA)*: 20 g glucose (or dextrose), 2 g $NaNO_3$, 0.5 g KH_2PO_4, 0.5 g $MgSO_4$, 1 g yeast extract, 1 mL $FeSO_4$ (from a 1% aqueous solution), 20 g agar, to 1 L distilled H_2O. Autoclave and allow to cool, and then add 13 mL/L from dichloran stock solution (*see* **Note 8**).
9. *Rose Bengal–glycerine–urea medium (RbGU)*: 10 g glycerol, 1 g urea, 0.5 g L-alanine, 750 mg PCNB, 0.5 g rose Bengal, 15 g agar, to 1 L distilled H_2O (*see* **Note 5**).
10. *Potato dextrose agar (PDA)*: 39 g of commercial PDA powder to 1 L of distilled H_2O (*see* **Note 9**).
11. *Potato dextrose broth (PDB)*: 39 g of commercial PDB powder to 1 L of distilled H_2O (*see* **Note 9**).

2.2 DNA Extraction

1. 15 mL polypropylene Falcon conical tubes.
2. Cetyl trimethyl ammonium bromide (CTAB) nucleic acid extraction buffer: 2% (*w*/*v*) CTAB, 20 mM EDTA·Na_2, 1.4 M NaCl, and 100 mM Tris, pH 8.0.
3. Sodium dodecyl sulfate (SDS) or sodium lauryl sulfate (SLS) stock solution: 20% (*w*/*v*) in distilled H_2O.
4. Glass beads (0.5 mm).
5. 70% ethanol.
6. 95% ethanol.
7. Solution of chloroform–isoamyl alcohol (24:1, *v*/*v*).
8. DNA storage buffer: 1× Tris–EDTA (TE) buffer (10 mM Tris–HCl, pH 7.6, 1 mM EDTA·Na_2).

9. DNA extraction kit: several commercial kits are available for extracting genomic DNA from plants and fungi are available and can be used to extract DNA from *Fusarium* spp. (*see* **Note 10**).
10. Nanodrop and/or Qubit.

2.3 PCR Amplification

1. Thermal cyclers.
2. Forward and reverse primers.
3. PCR master mix.
4. dNTPs.
5. DNA polymerase.

2.4 PCR-RFLP

1. Thermal cyclers.
2. Forward and reverse primers.
3. dNTPs.
4. DNA polymerase.
5. Restriction endonucleases.

2.5 Gel Electrophoresis

1. Gel electrophoresis instrument with gel molds and power supply.
2. Agarose powder.
3. Nucleic acid stain.
4. DNA ladder.
5. 1× TAE—prepare a 50× stock solution in 1 L of distilled H_2O: 242 g of Tris base, 57.1 mL of acetic acid (glacial), 100 mL of 0.5 M EDTA (pH 8.0). The 1× TAE solution is 20 mL of 50× stock in 980 mL distilled H_2O.

2.6 qPCR Reagents

1. Real-time thermal cycler (96- or 384-well format).
2. PCR plates (96- or 384-well plates).
3. Adhesive plate seals.
4. SYBR Green master mix: buffer, dNTPs, DNA polymerase, and SYBR Green dye.
5. If using TaqMan instead of SYBR green, then in addition to the master mix (buffer, dNTPs, DNA polymerase), species-specific primer/probe assay(s) is required.
6. Ultrapure nuclease-free water.

2.7 Software and Databases

1. *Primer3Plus*: a widely used online program for designing PCR primers.
   https://www.bioinformatics.nl/cgi-bin/primer3plus/primer3plus.cgi

2. *OligoAnalyzer*: a tool to determine many of the physical characteristics of your oligonucleotides.
   https://www.idtdna.com/pages/tools/oligoanalyzer?returnurl=%2Fcalc%2Fanalyzer
3. *NEBcutter*: a tool to find the restriction sites within DNA sequence
   https://nc2.neb.com/NEBcutter2/
4. *PrimerQuest*: a tool to design qPCR assays (primers and probes) for a wide variety of applications.
   https://www.idtdna.com/pages/tools/primerquest
5. *Fusarium_ID*: a publicly available database designed to support the identification of *Fusarium* spp. using sequences of multiple phylogenetically informative loci.
   http://isolate.fusariumdb.org/
6. *FUSARIUM MLST*: MultiLocus Sequence Typing to search for all species available in the current *Fusarium* database.
   https://fusarium.mycobank.org/

3 Methods

3.1 Recovering Fusarium spp. from Different Plant Tissues

Fusarium spp. can be associated with various parts of diseased plants including roots, stems, and grain/seeds. It is extremely important to differentiate between the *Fusarium* spp. that are the actual causal agent of the disease and the secondary pathogens that colonize the diseased tissue after infection. Working with freshly collected plant materials and using the appropriate isolation protocol with specific media act to reduce the number of secondary pathogens and saprophytes. Several agar-based media have been developed to isolate *Fusarium* spp. from different plant tissues (*see* details in Media Subheading 2.1).

1. Wash whole roots under running tap water for 15 min in a sieve to remove soil particles, and then cut into 1–2 cm pieces.
2. Root pieces are surface-disinfected with 0.5–1% sodium hypochlorite (NaOCl) for 2–3 min, then rinsed twice using sterile distilled water, and air-dried on sterilized filter papers.
3. Dried root pieces are then placed on *Fusarium*-specific medium, in 9 cm Petri dishes amended with antibiotics (*see* **Note 11**).
4. The plates are incubated for 2–5 days at 25 ± 2 °C and 12-h light/12-h dark. Growing hyphal tips were then transferred to PDA plates and incubated under the same condition (2–5 days at 25 ± 2 °C and 12-h light/12-h dark) to generate biomass for DNA extraction.

5. The same protocol used to recover *Fusarium* spp. from root tissue can be used to recover *Fusarium* spp. from stem and grain tissues. For the stem, lower stem pieces (node with ~1 cm segments on both sides) can be used. For grain, kernels with visual symptoms such as thin or shrunken and chalklike, sometimes with white or pinkish mold, are known collectively as *Fusarium*-damaged kernels (FDKs). These kernels are a good source to isolate *Fusarium* spp. involved in FHB (*see* **Note 12**).

3.2 DNA Extraction

Almost all *Fusarium* spp. are easy to cultivate using standard microbiological techniques, and for the nonexpert they can be obtained from various culture collections (*see* **Note 13**).

3.2.1 Extracting DNA from Pure Fusarium Cultures

1. For routine culturing, pure single-spore *Fusarium* isolates are maintained in Petri plates containing PDA medium.
2. From 5- to 7-day-old cultures, remove small agar plugs with growth and inoculate 250 mL Erlenmeyer flasks containing 50 mL PDB medium to generate biomass for DNA extraction. *Fusarium* spp. can produce large masses of spores (macroconidia and/or microconidia). Spore suspensions can be generated by growing *Fusarium* on water agar medium under light for 10 days, then adding 5–10 mL of sterile water to the agar plate, and gently shaking the plates to dislodge the spores. Inoculate the PDB media with the spore suspension.
3. Incubate liquid cultures at 20–25 °C for 3–7 days (temperature and days of incubation will vary among different *Fusarium* spp.) to generate biomass for DNA extractions.
4. Extract fungal whole cell DNA by filtering the cultures through a sterile Whatman # 1 filter paper.
5. Add 100–200 mg harvested mycelium to a sterile 15 mL Falcon polypropylene conical tube, and add 3 mL of CTAB nucleic acid extraction buffer plus 4 g of acid-washed and dry-baked 0.5 mm glass beads.
6. Vortex the mixture for 2–3 min, and add an additional 3 mL of lysis buffer to each tube.
7. Add ~660 μL of 20% SDS or SLS solution to a final concentration of 1% along with NaCl (1 M final concentration) and CTAB (1% final concentration). Mix tubes gently and incubate for a minimum of 1–2 h at 55 °C.
8. Extract cell debris, glass beads, denatured proteins, and lipids in 7 mL of chloroform–isoamyl alcohol (24:1, v/v) and pellet by centrifugation at 2500 rpm for 20 min at room temperature. Transfer the top aqueous layer to a 15 mL Falcon conical tube and precipitate the DNA by adding 2.5 volumes of ice-cold 95% ethanol.

9. Store tubes for about 3 h (*see* **Note 14**) at −20 °C and recover the DNA by centrifugation (1500× *g* for 30 min). Wash resulting pellets with 1 mL of 70% ethanol to remove excess salts. Air-dry the pellets and resuspend in 300 μL of 1× TE buffer.
10. You can replace the chloroform-extraction/ethanol-precipitation method described above with one of the commercially available DNA extraction kits (*see* **Note 10**) which will save time and effort. In this case, conduct **steps 1–4** from Subheading 3.2.1, then grind the mycelium in liquid nitrogen, and use the generated biomass powder as starting material to get DNA using any commercial DNA extraction kit according to manufacturer's recommendations. Alternately, mycelia can be freeze-dried (lyophilized) for 48–72 h and then ground at room temperature using a bead beater, e.g., Qiagen TissueLyzer II.

3.2.2 Extracting DNA from Infected Plant Tissues

1. Infected plant materials from cereal and/or pulse crops like roots, stems, and grains can be collected for DNA extraction. Samples are taken from plants (washed to remove soil particles for root samples) and subjected to surface disinfection with 0.5–2% sodium hypochlorite (NaOCl) for 2–3 min, then rinsed twice using sterile distilled water, and air-dried on sterilized filter papers.
2. Dried samples are then ground in liquid nitrogen to prepare fine powders that are used for DNA extraction. Usually, wheat node samples are very dry/fibrous and difficult to homogenize, so sterile fine sand (40–100 mesh) could be added during the grinding to facilitate homogenization.
3. Alternately, root and stem tissue can be flash frozen in liquid nitrogen, stored at −80°C and then freeze-dried for 48 h. Freeze-dried tissues are then ground in a bead beater at 1800 rpm using one 3 mm tungsten carbide bead in 2 mL tubes.
4. DNA can be extracted either with CTAB method (*see* Subheading 3.2.1) or with any commercial DNA extraction kit (*see* **Note 10**).
5. Quantify DNA on a spectrophotometer such as the Nanodrop or Qubit (*see* **Note 15**).

3.3 Detection Methods

Molecular techniques are a fast, reliable, and reproducible way to identify fungal pathogens including species within the genus *Fusarium*. Among the most popular of these methods are polymerase chain reaction (PCR), PCR-RFLP (restriction fragment length polymorphism), and real-time quantitative PCR (qPCR). Identifying different species of *Fusarium* morphologically and microscopically requires specialization and is laborious. These techniques will save time, money, and labor.

3.3.1 PCR Amplification

Polymerase chain reaction to amplify specific DNA fragments using primer pairs is an essential tool in *Fusarium* research. The *TEF1α* gene is preferred for species-specific identification [20–27], while the ITS region is targeted as a universal fungi barcode, but is not very informative for distinguishing *Fusarium* spp. [27, 28]. RPB1 and RPB2 are found as single copy genes and have a slow rate of sequence divergence and are especially useful in *Fusarium* for species identification [18]. Currently, two specific *Fusarium* databases are available: FUSARIUM-ID [19] and FUSARIUM MLST [29]. Both databases use the *TEF1α* gene to identify *Fusarium* spp. and the gene has been considered the sequence of choice to identify *Fusarium* isolates to species level [19, 29].

Several protocols were already published for specific detection of *Fusarium* spp. Table 1 contains selected references [20, 21, 23, 30–38] that provide species-specific primers for many phytopathogenic *Fusarium* spp. associated with cereal and/or pulse crops. We included a set of *TEF1α*-based primers [20, 21, 39] that can be used to detect and differentiate between *Fusarium* spp. associated with cereal and pulse crops (Fig. 2 and Table 2). Below, we provide a step-by-step general PCR protocol that can be used with different *Fusarium* spp., and specific primers for *Fusarium* spp. of interest can be selected from Tables 1 and 2, accordingly.

1. Select the suitable species-specific primer pair from Tables 1 and 2 depending on which *Fusarium* spp. you want to detect, and then follow the conditions outlined below.
2. Prepare PCR master mix in nuclease-free water: 1× PCR buffer, 200 μM dNTPs, 0.2 μM forward and reverse primers, 20–50 ng/μL template DNA (extracted as described in Subheading 2.2), and 1.25 units of Taq DNA polymerase per 50 μL reaction (*see* **Note 16**).
3. Set up PCR conditions: initial denaturation step at 94 °C for 3 min; followed by 25–35 cycles of denaturation (at 94 °C for 60 s), annealing (at appropriate T_a [annealing temperature; *see* **Note 17**] according to primers selected for 30–60 s), extension (at 72 °C for 1 min/1 Kb DNA amplicon); and a final extension step at 72 °C for 5–10 min.

3.3.2 Multiplex PCR

In multiplex PCR, several *Fusarium* spp. can be simultaneously detected in a single PCR reaction well, with a different pair of primers specific for each target *Fusarium* spp. Reagent master mix and PCR conditions can be conducted as explained in the PCR amplification section. Annealing temperature should be adjusted to work with all primer pairs used in multiplex PCR. Table 2 contains some references for detecting multiple *Fusarium* species using multiplex PCR with information about PCR conditions (i.e., annealing temperature).

Table 1
List of commonly used species-specific primers to detect *Fusarium* spp. associated with cereal and pulse crops

Primer	Sequence (5′–3′)	Target *Fusarium* spp.	Target gene/region	Band size (bp)	Reference
EF1 EF2	ATGGGTAAGGARGACAAGAC GGARGTACCAGTSATCATGTT	*Fusarium—universal*	*TEF1α*	700	[21]
Fg16F Fg16R	CTCCGGATATGTTGCGTCAA GGTAGGTATCCGACATGGCAA	*F. graminearum*	Fg16	400–500	[30]
UBC 85F UBC85R	GCAGGGTTTGAATCCGAGAC AGAATGGAGCTACCAACGGC	*F. graminearum*	SCAR85	332	[31]
FusF FgssR	TGGGTAARGAGGASAAGACTCACC TATGAGCCCCACCGGG	*F. graminearum*	*TEF1α*	368	[20]
FP82F FP82R	CAAGCAAACAGGCTCTTCACC TGTTCCACCTCAGTGACAGGTT	*F. poae*	ISSR	220	[23]
FAF1 FAR	AACATACCTTAATGTTGCCTCGG ATCCCCAACACCAAACCCGAG	*F. avenaceum*	ITS	413	[32]
FC01F FC01R	ATGGTGAACTCGTCGTGGC CCCTTCTTACGCCAATCTCG	*F. culmorum*	NA	570	[30]
CRO-AF CRO-AR	CTCAGTGTCCACCGCGTTGCGTAG CTCAGTGTCCCAATCAAATAGTCC	*F. cerealis*	NA	842	[33]
AF330109CF AF330109CR	AAAAGCCCAAATTGCTGATG TGGCATGTTCATTGTCACCT	*F. sporotrichioides*	*Tri13*	332	[34]
FsporF LansporR	CGCACAACGCAAACTCATC TACAAGAAGAGCGTGGCGATAT	*F. sporotrichioides*	*Tri5*	310	[35]

Redolens-F Redolens-R	ATCGATTTTCCCTTCGACTC CAATGATGATTGTGATGAGAC	*F. redolens*	*TEF1α*	387	[36]
FOF1 FOR1	ACATACCACTTGTTGCCTCG CGCCAATCAATTTGAGGAACG	*F. oxysporum*	*ITS*	340	[32]
TEF-Fs4f TEF-Fs4r	ATCGGCCACGTCGACTCT GGCGTCTGTTGATTGTTAGC	*F. solani*	*TEF1α*	658	[37]
FP1-1 FP1-2	CGGGGTAGTTTCACATTTCYG GAGAATGTGATGASGACAATA	*F. pseudograminearum*	*TEF1α*	523	[38]
FEF1 FER1	CATACCTATACGTTGCCTCG TTACCAGTAACGAGGTGTATG	*F. equiseti*	*TEF1α*	389	[31]

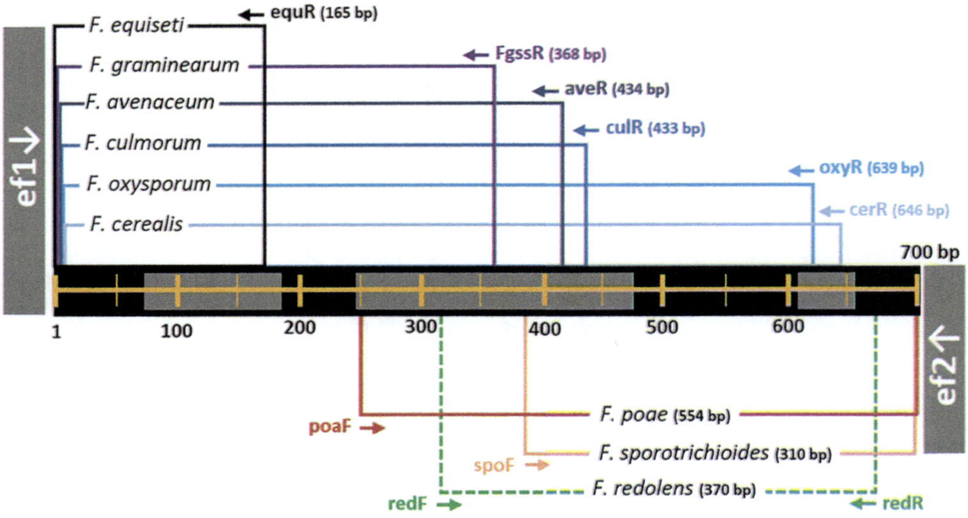

Fig. 2 Schematic representation of the *TEF1α* gene showing the exons (black boxes) and introns (gray boxes), with an internal scale (yellow lines) in relation to the region amplified by the *Fusarium-TEF1α* universal forward (ef1) and reverse (ef2) primers [21]. A set of species-specific primers and binding sites for nine *Fusarium* spp. associated with cereal and pulse crops are indicated [39]. *Fusarium-TEF1α* universal forward primer (ef1) can be paired with species-specific reverse primers to specifically amplify the target *Fusarium* species. Same with *Fusarium-TEF1α* universal reverse primer (ef2) that can paired with species-specific forward primers. A species-specific primer pair (redF/redR) is used to specifically detect *Fusarium redolens*. The expected band length for each species is indicated between parentheses. (Information about these primers are listed in Table 2)

Table 2
List of species-specific primers based on the *TEF1α* gene sequence to detect *Fusarium* spp. commonly associated with cereal and pulse crops

Primer	Direction	Sequence (5′–3′)	Target species	T_m °C	Reference
efl	Forward	ATGGGTAAGGARGACAAGAC	*Fusarium* spp.	52.7	[21]
ef2	Reverse	GGARGTACCAGTSATCATGTT	*Fusarium* spp.	53.1	[21]
FgssR	Reverse	TATGAGCCCCACCGGG	*F. graminearum*	57.1	[20]
aveR	Reverse	ATAGTGCGCGACAGTGTG	*F. avenaceum*	53.3	[39]
cerR	Reverse	TCGTGTTAGTATGAGAAGGTC	*F. cerealis*	51.3	[39]
equR	Reverse	AATCTTGTCAGCAAATGTATARG	*F. equiseti*	50.0	[39]
oxyR	Reverse	AGAGAAGTAGAATGAAGCATG	*F. oxysporum*	49.7	[39]
poaF	Forward	GGTAGACTCAACATGCACT	*F. poae*	51.9	[39]
spoF	Forward	CTTCCCACACATCCATTTACA	*F. sporotrichioides*	52.7	[39]
redF	Forward	GACTCGCCGCTCCCAT	*F. redolens*	57.9	[39]
redR	Reverse	CAATGATGATTGTGATGAGACATTAGC	*F. redolens*	54.5	[39]

Melting temperature (T_m °C) is indicated for each primer. See Fig. 1 for more details about binding sites and expected band length

3.3.3 Gel Electrophoresis

1. To prepare 1.5% agarose gel, add 1.5 g of agarose to a flask.
2. Add 100 mL of 1× TAE running buffer.
3. Mix until agarose is incorporated into the buffer.
4. Use a microwave to dissolve the agarose into the buffer by setting on high for 2 min (*see* **Note 18**).
5. Add nucleic acid stain (*see* **Note 19**) to the agarose when it is cooler (60 °C) and swirl to dissolve.
6. While waiting for the agarose solution to cool, prepare gel mold and comb as required.
7. When agarose has cooled to between 50 °C and 60 °C, pour into the mold containing the comb.
8. Allow gel to set for approximately 30 min, or until a gloved finger gently applied to the lower right corner feels firm and looks opaque.
9. Remove comb and place gel that is in the mold into the electrophoresis instrument.
10. Add a DNA ladder to the far left well, and right if required, and then add PCR products.
11. Run the electrophoresis instrument at between 80 and 120 V for 1–2 h in 1× TAE buffer, or until the tracking dye reaches the bottom of the gel.
12. Remove the gel from the instrument and visualize the bands using UV light.

3.3.4 PCR-RFLP

The PCR restriction fragment length polymorphism (PCR-RFLP) is considered an easy, rapid, and accurate technique for fungal identification. It is a method based on the digestion of PCR amplicons with appropriate restriction enzymes to produce distinct polymorphic fragments used as markers for species identification. Several PCR-RFLP protocols have been published to differentiate between phytopathogenic *Fusarium* spp. Table 3 provides references [20, 25, 36, 39–47] for using PCR-RFLP to differentiate between *Fusarium* spp. associated with cereal and pulse crops.

A large number of restriction enzymes are commercially available and many online tools (such as NEBcutter) are available that will accept an input DNA sequence and produce a comprehensive report of the restriction enzymes that will cleave the sequence.

1. Perform PCR using species-specific primers as described in Subheading 3.3.1 and confirm amplicon size by agarose gel electrophoresis (*see* Subheading 3.3.3).
2. Conduct restriction digestion for the PCR amplicons according to **steps 5–9**.

Table 3
PCR-RFLP protocols to differentiate between *Fusarium* spp. commonly associated with cereal and pulse crops

Reference	Gene/ DNA region	*F. acuminatum*	*F. avenaceum*	*F. cerealis*	*F. compactum*	*F. culmorum*	*F. equiseti*	*F. graminearum*	*F. langsethiae*	*F. oxysporum*	*F. poae*	*F. pseudograminearum*	*F. redolens*	*F. semitectum*	*F. solani*	*F. sporotrichioides*	*F. tricinctum*	FGSC
41	IGS									✓								
37	ITS									✓			✓					
42	ITS									✓								
43	*CTP51C*		✓														✓	
44	*TEF1α*																	✓
20	*TEF1α*																	✓
40	*TEF1α*		✓	✓		✓	✓	✓		✓	✓		✓			✓		
45	ITS	✓			✓		✓					✓		✓	✓			
46	*TEF1α/ β-tub*									✓								
47	IGS								✓		✓					✓		
48	ITS			✓		✓		✓		✓	✓							
49	*β-tub*			✓		✓		✓										

✓: indicate *Fusarium* spp. that can be differentiated from others and produce unique restriction pattern when subjected to the PCR-RFLP protocol as descried in corresponding reference.

FGSC: *Fusarium graminearum* species complex.

3. Use 400 ng of the PCR product (usually 10 μL of PCR is enough).
4. Use 10–15 U of a suitable restriction endonuclease (see Table 3 for endonucleases used to differentiate between *Fusarium* spp. associated with cereal and pulse crops).
5. Add 2 μL of the endonuclease buffer.
6. Adjust volume to 30 μL with nuclease-free water.
7. Incubate at 37 °C for 5–10 min.
8. Inactivate restriction endonuclease by heating at 60–70 °C for 15 min.
9. Run 2% agarose gel (*see* Subheading 3.3.3) using the entire reaction volume (30 μL) and visualize the bands using UV light.

3.3.5 qPCR Quantification

In quantitative analysis, two methods are preferred: SYBR Green and TaqMan. SYBR Green is a method that works by intercalating the minor grooves of any dsDNA (nonspecific). As dsDNA is amplified, an increase in signal results. TaqMan is a sequence-specific method based on the hydrolysis probe. It works by using a reporter-quencher combination. As DNA polymerase synthesizes new strands of DNA, it cleaves the reporter allowing a signal to be emitted which increases as more DNA is amplified. Both technologies are designed to generate fluorescence during the PCR, which allows the qPCR machine to monitor the reaction in "real time." Generally, if running a few samples and only singleplex, SYBR Green can save money, but if multiplex and high throughput are required, then TaqMan is mandatory since the probe is species-specific.

3.4 Quantify Fusarium spp. in Cereal and/or Pulse Host Tissues

3.4.1 Master Mix Preparation

1. *TEF1α*-based TaqMan qPCR assays [8, 22] that can be used to detect and quantify important *Fusarium* spp. associated with wheat and/or pulse crops are listed in Table 4.
2. Use the calculation: 10 μL of qPCR master mix; 1 μL of species-specific qPCR assay (choose from Table 4 a specific assay to your target *Fusarium* spp.) and 5 μL DNase/RNase-free water for each reaction.
3. Multiply these amounts by the total number of reactions you will use.
4. Mix the prepared master mix with all components (except DNA), and dispense 16 μL/well.
5. Add 4 μL normalized DNA to each well separately, and design the experiment to measure each DNA samples three times (triplicates). The qPCR instrument will measure the standard deviation and calculate the average DNA amount for the three replicates.
6. Include controls: positive control is a pure DNA sample from the *Fusarium* species being quantified with known concentration, and negative control is DNase/RNase-free water. Use 4 μL/reaction for both positive and negative controls and include three replicates each.
7. Cover with only qPCR seal film, without touching the film. Line up and press edges only; use Kim wipes to smooth the film tight; spin down the plate to remove bubbles.
8. Prepare a tenfold serial dilution of DNA from each *Fusarium* species that will be assayed in the reaction (six data points starting from at 1:10 down to 0.0001 ng/μL).
9. Prepare a tenfold serial dilution using pure DNA from the host plant for the standard curve calculations (*see* **Note 20**).

Table 4
TaqMan qPCR assays designed to target species-specific regions within the *TEF1α* gene of major *Fusarium* spp. associated with cereal and pulse crops

Species	Primer/probe	Sequence (5′–3′)	Reference
F. avenaceum	qFavF	ACCACTGTAAGTACAACCATCAGCGAGTC	[8]
	qFavR	CGGTCTGTCAAGAGTTAGCAAGATGTCG	
	qFavP	**FAM**/TCTGCACTC/**ZEN**/ GGAACCCGCCAAACCTG/**31ABkFQ**	
F. culmorum	qFcuF	ATTTTGCGGCTTTGTCGTAATTTTTCTG	[8]
	qFcuR	TGACACGTGATGGTGCGCGCCT	
	qFcuP	**HEX**/CAGGCGCTT/**ZEN**/ GCCCTCTTCCCACAAACCA/**3IABkFQ**	
F. graminearum	qFgrF	TGCGGCTTTGTCGTAATTTTTTYCCC	[8]
	qFgrR	AGTGACTGGTTGACACGTGATGATGA	
	qFgrP	**FAM**/CAGGCGTCT/**ZEN**/ GCCCTCTTCCCACAAACCA/**3IABkFQ**	
F. poae	qFpoF	GCGGGGTAGACTCAACATGCACT	[8]
	qFpoR	ATTCGAGTGATGGATCGAGGGAAAGT	
	qFpoP	**HEX**/ATGCTTGAC/**ZEN**/ AGACCGGTCACTTGATCATCCAGTG/ **3IABkFQ**	
F. acuminatum	AcuF	TGGCAGGGTATCACCAAAA	[22]
	AcuR	CGGGGTAATGGATCTGTTTC	
	AcuP	**CAL Fluor Red 610**/ CCACGATTCGCTCCCTCACT/**BHQ-2**	
F. redolens	RedF	CCCTCTTCCCACACAATCAC	[22]
	RedR	AGCTCAGCGGCTTCCTATTA	
	RedP	**FAM**/TGAGCGCGGATCATCACGTG/**BHQ-1**	
F. solani	SolF	GCGCCTTACTATCCCACATC	[22]
	SolR	TTTTGTGACTCGGGAGAAGC	
	SolP	**HEX**/CCTCCTCCGCGACACGCTCT/**BHQ-1**	
F. sporotrichioides	SpoF	TTTTTACGGCTGTGTCGTGA	[22]
	SpoR	CGGCTTCCTATTGACAGGTG	
	SpoP	**CAL Fluor Red 610**/ TGATAGTGGGGCTCATACCC/**BHQ-2**	

TaqMan probes can be labelled with a reporter dye (CAL, FAM or HEX) on the 5′-end. Probes also contained the Black Hole Quencher at the 3′-end with the additional internal quencher ZEN in the middle of some probes
F forward primer, *R* reverse primer, *P* probe

3.4.2 Cycling Conditions and Obtaining Results

1. Initial denaturation at 95 °C for 3 min.
2. Followed by 40 cycles of denaturation at 95 °C for 5 s and annealing/extension at 60 °C for 30 s.
3. Analyze the wells within the software according to the manufacturer's directions.
4. Export a copy of the results (typically to Excel) for analysis.

4 Notes

1. Media is autoclaved and allowed to cool (~50 °C) before adding the antibiotic stock solutions. Antibiotic stock solutions are sterilized by filtration (using 0.45 μm filters). Streptomycin is effective against Gram-negative bacteria, and neomycin against Gram-positive bacteria. Adding antibiotics to media is recommended when isolating *Fusarium* spp. from plant tissues to suppress bacterial growth. Adding antibiotics to media when working with pure *Fusarium* cultures to generate biomass for DNA extraction is optional.
2. Adding 4–8 g KCl to WA will help the fungal cultures produce more and longer microconidial chains.
3. Usually, one carnation leaf piece is added per 2 mL of medium (~10 pieces per 9 cm diameter Petri dish). Can use irradiated banana leaf pieces if carnation leaves are not available. Visit https://wiki.bugwood.org/Carnation_leaf_agar for more information about carnation leaf agar.
4. Place 1–2 pieces of sterile filter paper (Whatman #1; ~1 cm^2) on the agar surface after the medium has solidified to increase sporulation.
5. PCNB is usually added as 1 g of Terrachlor, which contains 75% PCNB (*w/w*). If Terrachlor is not available, pure PCNB (Sigma Aldrich) can be used, and although it has incomplete solubility in water, semi-selective properties are retained.
6. Preferred to use Fe3Na EDTA when the target is *F. oxysporum*, which appear distinctly pigmented on this medium, and usually separable from other *Fusarium* spp.
7. The only difference between MGA and NSM/PPA is that the PCNB is replaced by 2.5 ppm malachite green due to the carcinogenicity of PCNB.
8. Dichloran (2,6-dichloro-4-nitroanaline) is prepared as 50 mg of dichloran in 100 mL ethanol and used at the rate of 13 mL/L SFA. If dichloran is not available, then PCNB, as used in PPA, can be incorporated into the medium prior to autoclaving.
9. Commercially, PDA and PDB are available from multiple suppliers offering virtually the same product.
10. The most commonly used kits for fungal DNA extraction from plant materials such as DNeasy Plant Mini Kit (Qiagen), DNeasy Plant Pro Kit (Qiagen), GenElute Plant Genomic DNA Miniprep Kit (MilliporeSigma), Norgen's Plant/Fungi DNA Isolation Kit (Millipore-Sigma), and many other kits are available.

11. With soilborne diseases like root rot in cereal and pulse crops, you should expect a range of fungi other than the original disease causal agent. In this case, *Fusarium*-specific media could be used to favor the growth of *Fusarium* spp. and suppress other genera. NSM supplemented with fungicide (PCNB) is highly inhibitory to most other fungi and bacteria but allows slow growth of *Fusarium*. MGA could also be used to isolate *Fusarium* spp. from plant roots and inhibit other common root-associated fungal communities like *Aspergillus*, *Penicillium*, and Zygomycetes without reducing the number of *Fusarium* spp. MGA medium was developed as an alternative of NSM in which PCNB is replaced by malachite green. SFA medium could also be used for selective isolation of *Fusarium* spp. from plant debris and roots. The medium permits for the slow growth of *Fusarium* spp. from root fragments and inhibit other fungal genera.
12. The *Fusarium* spp. predominantly found associated with root rot of wheat and other small-grain cereals (e.g., barley and oat) are *F. solani* and *F. oxysporum*. However, some other *Fusarium* spp. were recently found to be associated with Fusarium root rot (FRR) including *F. graminearum*, *F. culmorum*, and *F. cerealis*. The *Fusarium* spp. predominantly found associated with *Fusarium* crown/stem rot (FCR) are *F. culmorum* and *F. pseudograminearum*. The *Fusarium* spp. predominantly found associated with *Fusarium* head blight (FHB) in wheat and other small-grain cereals are *F. graminearum*, *F. poae*, *F. avenaceum*, and *F. culmorum*.
13. Culture collections: **CBS** = Centraalbureau voor Schimmelcultures, Utrecht, the Netherlands; **ATCC** = American Type Culture Collection, Manassas, VA, USA; **ARS** (**NRRL**) = culture collection of Agricultural Research Service, United States Department of Agriculture, IL, USA; **DAOM** = Department of Agriculture, Ottawa, Mycology/Canadian National Mycological Herbarium, Ottawa, ON, Canada; and **UAMH** Centre for Global Microfungal Biodiversity, Toronto, ON, Canada. Also keep in mind that suitable import permits may have to be obtained for ordering and importing microorganisms.
14. Ice-cold 95% ethanol tubes from this step could be stored at −20 °C overnight, and perform remaining steps the next day.
15. Normalize extracted DNA samples using the equation $C_1 V_1 = C_2 V_2$. This step is important to correct inter-sample variations
16. Add the Taq DNA polymerase last so that the reaction does not start prematurely and efficacy is maintained by keeping it at −20 °C until needed. Subpar results or no amplification can occur if Taq is added too soon to the reaction mixture.

17. Annealing temperature (T_a): the annealing temperature is determined by the melting temperature (T_m) of the selected primers for PCR amplification. A general rule of thumb is to begin with an annealing temperature 3–5 °C lower than the lowest T_m of the primers. Many tools are available online to calculate the T_m of primers and estimates an appropriate annealing temperature (*see* "Software and databases" Subheading 2.7).
18. Use a 500 mL or larger flask to avoid having the agarose/buffer mixture overflow resulting in a messy microwave and a loss of gel solution, making the gel smaller than it should be. For larger volumes, for example, 200 mL, add 3 g of agarose and 200 mL of 1× TBE buffer. Microwave on high for 3 min. You will know when the solution is done when there are no crystals floating in it; it should be completely clear.
19. Ethidium bromide (EtBr) is the classic DNA stain dye. However, due to its toxicity (i.e., mutagen), it is recommended that EtBr be replaced by safer alternatives, such as SYBR-Safe (manufactured by Invitrogen), GelRed, or GelGreen (manufactured by Biotium).
20. Limit of detection (LOD) and limit of quantification (LOQ) for each assay can be determined by running seven-point standard curves (10 ng to 10^{-5} ng), with 20 "unknown" replicates for the lowest three concentrations. LOD is the concentration at which a single detection failure out of 20 replicates is recorded (95% confidence), and LOQ is when the standard deviation of the threshold cycle (Ct) is less than 0.5.

References

1. Munkvold GP (2017) *Fusarium* species and their associated mycotoxins. In: Moretti A, Susca A (eds) Mycotoxigenic fungi. Methods in molecular biology, vol 1542. Humana Press, New York, pp 51–106
2. Chatterton S, Harding MW, Bowness R et al (2019) Importance and causal agents of root rot on field pea and lentil on the Canadian prairies 2014–2017. Can J Plant Pathol 41: 98–114
3. Chittem K, Mathew F, Gregoire M et al (2015) Identification and characterization of *Fusarium* spp. associated with root rots of field pea in North Dakota. Eur J Plant Pathol 143:641–649
4. Gossen B, Conner R, Chang KF et al (2016) Identifying and managing root rot of pulses on the northern Great Plains. Plant Dis 100: 1965–1978
5. Safarieskandari S, Chatterton S, Hall LM (2021) Pathogenicity and host range of *Fusarium* species associated with pea root rot in Alberta, Canada. Can J Plant Pathol 43:162–171
6. Moparthi S, Burrows M, Mgbechi-Ezeri J, Agindotan B (2021) *Fusarium* spp. associated with root rot of pulse crops and their cross-pathogenicity to cereal crops in Montana. Plant Dis 105:548–557
7. Bai G, Shaner G (2004) management and resistance in wheat and barley to Fusarium head blight. Annu Rev Phytopathol 42:135–161
8. Hafez M, Gourlie R, Telfer M et al (2021) Diversity of *Fusarium* spp. associated with wheat node and grain in representative sites across the western Canadian Prairies. Phytopathology 112:1003–1015

9. Osborne LE, Stein JM (2007) Epidemiology of Fusarium head blight on small-grain cereals. Int J Food Microbiol 119:103–108
10. Aboukhaddour R, Fetch T, McCallum BD et al (2020) Wheat diseases on the prairies: a Canadian story. Plant Pathol 69:418–432
11. Aoki T, O'Donnell K, Geiser DM (2014) Systematics of key phytopathogenic *Fusarium* species: current status and future challenges. J Gen Plant Pathol 80:189–201
12. Scherm B, Balmas V, Spanu F et al (2013) *Fusarium culmorum*: causal agent of foot and root rot and head blight on wheat. Mol Plant Pathol 14:323–341
13. Zeilinger S, Gupta VK, Dahms TE et al (2016) Friends or foes? Emerging insights from fungal interactions with plants. FEMS Microbil Rev 40:182–207
14. Karim NFA, Mohd M, Nor NMIM, Zakaria L (2016) Saprophytic and potentially pathogenic *Fusarium* species from peat soil in Perak and Pahang. Trop Life Sci Res 27:1
15. Edel V, Steinberg C, Gautheron N, Alabouvette CJMR (1997) Evaluation of restriction analysis of polymerase chain reaction (PCR)-amplified ribosomal DNA for the identification of *Fusarium* species. Mycol Res 101:179–187
16. Kashyap PL, Rai S, Kumar S et al (2015) Mating type genes and genetic markers to decipher intraspecific variability among *Fusarium udum* isolates from pigeonpea. J Basic Microbiol 55: 846–856
17. Mach RL, Kullnig-Gradinger CM, Farnleitner et al (2004) Specific detection of *Fusarium langsethiae* and related species by DGGE and ARMS-PCR of a β-tubulin (tub1) gene fragment. Int J Food Microbiol 95:333–339
18. O'Donnell K, Rooney AP, Proctor RH et al (2013) Phylogenetic analyses of RPB1 and RPB2 support a middle Cretaceous origin for a clade comprising all agriculturally and medically important fusaria. Fungal Genet Biol 52:20–31
19. Geiser DM, del Mar J-GM, Kang S et al (2004) FUSARIUM-ID v. 1.0: a DNA sequence database for identifying *Fusarium*. Eur J Plant Pathol 110:473–479
20. Hafez M, Abdelmagid A, Adam LR, Daayf F (2020) Specific detection and identification of *Fusarium graminearum* Sensu Stricto using a PCR-RFLP tool and specific primers targeting the translational elongation factor 1α gene. Plant Dis 104:1076–1086
21. O'Donnell K, Kistler HC, Cigelnik E, Ploetz RC (1998) Multiple evolutionary origins of the fungus causing Panama disease of banana: concordant evidence from nuclear and mitochondrial gene genealogies. Proc Natl Acad Sci U S A 95:2044–2049
22. Zitnick-Anderson K, Simons K, Pasche JS (2018) Detection and qPCR quantification of seven *Fusarium* species associated with the root rot complex in field pea. Can J Plant Pathol 40: 261–271
23. Parry DW, Nicholson P (1996) Development of a PCR assay to detect *Fusarium poae* in wheat. Plant Pathol 45:383–391
24. Nicolaisen M, Suproniené S, Nielsen LK et al (2009) Real-time PCR for quantification of eleven individual *Fusarium* species in cereals. J Microbiol Methods 76:234–240
25. Kachuei R, Yadegari MH, Safaie N et al (2015) PCR-RFLP patterns for the differentiation of the *Fusarium* species in virtue of ITS rDNA. Curr Med Mycol 1:4
26. O'Donnell K, Whitaker B, Laraba I et al (2022) DNA sequence-based identification of *Fusarium*: a work in progress. Plant Dis 106:1597–1609
27. Torres-Cruz T, Whitaker B, Proctor R et al (2022) FUSARIUM-ID v.3.0: an updated, downloadable resource for *Fusarium* species identification. Plant Dis 106:610–1616
28. Schoch CL, Seifert KA, Huhndorf S et al (2012) Nuclear ribosomal internal transcribed spacer (ITS) region as a universal DNA barcode marker for Fungi. Proc Natl Acad Sci U S A 109:6241–6246
29. O'Donnell K, Sutton DA, Rinaldi MG et al (2010) Internet-Accessible DNA sequence database for identifying fusaria from human and animal infections. J Clin Microbiol 48: 3708–3718
30. Nicholson P, Simpson DR, Weston G et al (1998) Detection and quantification of *Fusarium culmorum* and *Fusarium graminearum in* cereals using PCR assays. Physiol Mol Plant Pathol 53:17–37
31. Schilling AG, Moller EM, Geiger HH (1996) Polymerase chain reaction-based assays for species-specific detection of *Fusarium culmorum*, *F. graminearum*, and *F. avenaceum*. Phytopathology 86:515–522
32. Mishra PK, Fox RT, Culham A (2003) Development of a PCR-based assay for rapid and reliable identification of pathogenic Fusaria. FEMS Microbiol Lett 218:329–332
33. Yoder WT, Christianson LM (1997) RAPD-derived primers for distinguishing members of the section Fusarium. Cereal Res Commun 25: 571–575
34. Demeke T, Clear RM, Patrick SK, Gaba D (2005) Species-specific PCR-based assays for the detection of *Fusarium* species and a

comparison with the whole seed agar plate method and trichothecene analysis. Int J Food Microbiol 103:271–284

35. Wilson A, Simpson D, Chandler E et al (2004) Development of PCR assays for the detection and differentiation of *Fusarium sporotrichioides* and *Fusarium langsethiae*. FEMS Microbiol Lett 233:69–76
36. Bogale M, Wingfield BD, Wingfield MJ, Steenkamp ET (2007) Species-specific primers for *Fusarium redolens* and a PCR-RFLP technique to distinguish among three clades of *Fusarium oxysporum*. FEMS Microbiol Lett 271:27–32
37. Arif M, Chawla S, Zaidi MW et al (2012) Development of specific primers for genus *Fusarium* and *F. solani* using rDNA sub-unit and transcription elongation factor (TEF-1α) gene. Afr J Biotechnol 11:444–447
38. Aoki T, O'Donnell K (1999) Morphological and molecular characterization of *Fusarium pseudograminearum* sp. nov., formerly recognized as the Group 1 population of *F. graminearum*. Mycologia 91:597–609
39. Hafez M, Abdelmagid A, Aboukhaddour R et al (2021) Fusarium root rot complex in soybean: molecular characterization, trichothecene formation, and cross-pathogenicity. Phytopathology 111:2287–2302
40. Appel DJ, Gordon TR (1995) Intraspecific variation within populations of *Fusarium oxysporum* based on RFLP analysis of the intergenic spacer region of the rDNA. Exp Mycol 19:120–128
41. Dubey SC, Tripathi A, Singh SR (2010) ITS-RFLP fingerprinting and molecular marker for detection of *Fusarium oxysporum* f. sp. *ciceris*. Folia Microbiol 55:629–634
42. Fernández-Ortuño D, Atkins SL, Fraaije BA (2011) The use of a CYP51C gene based PCR-RFLP assay for simultaneous detection and identification of *Fusarium avenaceum* and *F. tricinctum* in wheat. Int J Food Microbiol 145:370–374
43. Garmendia G, Umpierrez-Failache M, Ward TJ, Vero S (2018) Development of a PCR-RFLP method based on the transcription elongation factor 1-α gene to differentiate *Fusarium graminearum* from other species within the *Fusarium graminearum* species complex. Food Microbiol 70:28–32
44. Kim JS, Kang NJ, Kwak YS, Lee C (2017) Investigation of genetic diversity of *Fusarium oxysporum* f. sp. *fragariae* using PCR-RFLP. Plant Pathol J 33:140–147
45. Konstantinova P, Yli-Mattila T (2004) IGS–RFLP analysis and development of molecular markers for identification of *Fusarium poae*, *Fusarium langsethiae*, *Fusarium sporotrichioides* and *Fusarium kyushuense*. Int J Food Microbiol 95:321–331
46. Llorens A, Hinojo MJ, Mateo R et al (2006) Characterization of *Fusarium* spp. isolates by PCR-RFLP analysis of the intergenic spacer region of the rRNA gene (rDNA). Int J Food Microbiol 106:297–306
47. Nosratabadi M, Kachuei R, Rezaie S et al (2018) Beta-tubulin gene in the differentiation of *Fusarium* species by PCR-RFLP analysis. Infez Med 26:52–60

Chapter 2

Distinguishing *Puccinia striiformis* f. sp. *tritici* Isolates Using Genomic Sequencing: A Case Study

Michele Frick, Eric Amundsen, and André Laroche

Abstract

We are reporting on the utilization of high-throughput sequencing and different sequencing analysis tools to delineate identification of different isolates of the stripe rust fungal pathogen *Puccinia striiformis* f. sp. *tritici* (*Pst*). Different approaches are shown: utilization of rDNA sequences and random sequences that may be very useful to make sure that isolates belong to *Pst* and to distinguished closely related isolates. Identification of unique/lost sequences could lead to the identification of effectors associated with specific isolates.

Key words Bioinformatics, DNA isolation, High throughput sequencing, Polymorphisms, *Puccinia striiformis* f. sp. *tritici*, Random sequences, rDNA sequences, Wheat pathogen

1 Introduction

Feeding a potential population of 9 billion by 2050 is a very important challenge of agriculture. Given the fact that plant pathogens can cause up to 30–40% yield losses, improving plant resistance against a myriad of different pathogens is a critical issue. An important survey toward identifying the 10 most important plant fungal pathogens placed the three wheat rusts as the third highest-ranking pathogens [1]. The three wheat rusts are stem rust caused by *Puccinia graminis* f. sp. *tritici* (*Pgt*), leaf rust caused by *P. triticina* (*Pt*), and stripe rust caused by *P. striiformis* f. sp. *tritici* (*Pst*). They are found everywhere wheat is grown around the world [2]. Their relative abundance depends on their location, temperature, and altitude. All three rusts are obligate biotrophic fungi that are highly specific to their host and all three have an heteroecious life cycle that implies the presence of a primary host for asexual reproduction (e.g., wheat) and an alternate host for sexual reproduction of the fungal pathogen [2]. An important tenet of biology is that genetic variation is usually introduced during meiosis during

Nora A. Foroud and Jonathan A. D. Neilson (eds.), *Plant-Pathogen Interactions*, Methods in Molecular Biology, vol. 2659, https://doi.org/10.1007/978-1-0716-3159-1_2,

the sexual cycle. Sexual reproduction leads to genetic recombination that increases variability of spores and their ability to defeat host resistance genes. This is why we saw an aggressive eradication of the alternate host, barberry, for stem rust in the first part on the twentieth century and its removal from the environment led to great diminution of the number of variants, and a long-time stability in the predominant resistance genes in Canada and around the world [3]. Epidemics of stem and leaf rusts in the middle of the twentieth century have been very important and lead to studies on the identification and incorporation of effective and lasting resistance genes into the wheat germplasm [4]. Recently, a high level of somatic variation was identified in *Pst* which would explain the ability of stripe rust to mutate rapidly and its ability to rapidly defeat wheat lines carrying a single resistance gene [5]. This was interesting as it took a very long time to identify the alternate host for *Pst* [6]. This somatic activity is much higher in *Pst* than in *Pgt* and *Pt*. Until the year 2000 in North America, *Pst* was considered the least threatening of the three rusts as it was found in limited geographic areas. However, since 2000 in both North America and around the world, more aggressive and more heat-tolerant stripe rust isolates have been identified [7–10] and *Pst* is now a serious threat to wheat production worldwide [11].

Consequently, the ability to distinguish *Pst* isolates is important to assess the variability and constitution of stripe rust samples. In this chapter, we are reporting on the classical way to characterize strains based on their complement of avirulence genes and their ability to defeat known individual stripe rust resistance gene. But more importantly, we provide information on our protocols to characterize purified *Pst* isolates based on their genomic DNA content. The integration of the information provided is assembled into a case study.

2 Materials

2.1 Plant Rearing and Rust Isolate Inoculation and Collection

1. A susceptible wheat cultivar is needed to ensure that stripe rust infection is successful (*see* **Note 1**).
2. Stripe rust isolates: obtained from naturally infected wheat lines or from a collection. All isolates must have been pustule purified twice.
3. Growth cabinet or greenhouse to grow wheat plants.
4. Dew chamber to favor wheat infection (*see* **Note 2**).
5. Inoculation fluid: Novec 7100 Engineered Fluid [12] or Soltrol® 170 Isoparaffin [13] to inoculate stripe rust urediniospores to susceptible wheat (*see* **Note 3**) or alternatively, talc powder.

6. Double distilled water with nonionic detergent (1 drop/L) (*see* **Note 4**).
7. Spore inoculation: Use 10 mg spores/mL of inoculation fluid (1 mg of spores $\approx 2.45 \times 10^5$ spores).
8. A mini cyclone spore collector (*see* **Note 5**).
9. A spore inoculator (*see* **Note 6**).
10. Standard 00 gelatin capsule or 20 mL glass vial to attach to the spore inoculator or spore collector (*see* **Notes 5** and **6**).
11. 50 mL sterile tubes with screw cap.
12. Sterile 10 μL pipet tips.
13. Disposable gloves.
14. Plastic/cellophane bags to cover pots and trays of pots to maintain high humidity levels.
15. Water sprayer bottle (500 mL or 1 L).
16. 70% ethanol.
17. Compressed air source or compressor delivering pressure between 10 and 15 PSI (70–100 kP).

2.2 DNA Extraction from Stripe Rust Spores and Sequencing

1. A bead-beating grinding instrument (*see* **Note 7**).
2. 2 mL reinforced (tough) conical tubes (*see* **Note 8**).
3. Synthetic or glass beads to help break up the spores (*see* **Note 9**).
4. Diatomaceous earth is also needed to help break up the spores (*see* **Note 10**).
5. Liquid nitrogen.
6. Any DNA extraction kit or your favorite DNA extraction protocol compatible with DNA sequencing (*see* **Note 11**).
7. DNA sequencing: in-house or a commercial sequencing provider (*see* **Note 12**).

2.3 Software

1. Commercial sequencing analysis software are quick and easy to use for analyzing sequencing results and assembling data (*see* **Note 13**).
2. Alternatively, you can use open-source software. They are easily accessible but the learning curve can be challenging especially if your knowledge of "command line" is limited (*see* **Note 14**).

3 Methods

3.1 Identification of Stripe Rust Isolates

An initial step is the identification of infected wheat leaves by *Puccinia striiformis* f. sp. *tritici* (*Pst*), the causal agent of stripe rust in experimental wheat plots or in growers' field in your area of interest. Ditches next to growers' fields are another location to collect spores as they may escape fungicide treatments of fields. Where spring and winter wheat are coexisting, one can serve as a bridge to the other one, so usually, the "green crop" is the place to look to identify spores. Leaves exhibiting spores should be individually collected and stored in individual paper or plastic envelopes. Upon the return to the lab from the collection trip, individual well-spaced pustules are picked up and inoculated as below. Applying a given source of *Pst* spores to differential lines carrying different resistance genes is a way to initially characterize different isolates. Genomic sequencing of collected spores [14] is the approach that we have selected. Transcriptomic sequencing of infected leaf tissue [15] can be an alternative approach but will provide a composite report if more than one isolate is infecting a given single leaf.

3.2 Single Spore Isolation of Stripe Rust

1. Susceptible wheat plants (e.g., Avocet S; *see* **Note 1**) for inoculation are grown at 20 °C to the four-leaf (all stage resistance) to flag leaf (adult resistance) stages in a growth chamber or greenhouse.
2. Pour a small amount of inoculation fluid [12] or [13] into a sterile petri dish (*see* **Note 3**).
3. First, gently rub the leaf with glove fingers and then wet the gloved fingers with the inoculation solution and gently wet a larger leaf of the plant for inoculation placing the thumb and index finger on each side of the leaf.
4. Change gloves.
5. Select a single urediniospore from an infected leaf on the end of a 10 μL pipet tip and then put the tip down on a clean box so the end is not touching anything.
6. Change gloves.
7. Choose a larger leaf for inoculation. Inoculate leaf with the end of the tip about 1/3 of the way up the leaf.
8. Then rub from the inoculation point to the end of the leaf.
9. Change gloves.
10. Put a labelled stake in the soil beside the inoculated leaf or cut the end of the leaf tip just to mark it.
11. Allow to dry for 30 min to evaporate all oil (*see* **Note 15**).

12. Once dry, add a clean stake to the pot. Alternatively, mix spores with talc (1:20 if spores were previously frozen or 1:50 if spores were freshly collected) and apply the mixture to leaves with an autoclaved cotton swab.
13. Mist plants with water (containing 1 drop of nonionic detergent/L) (*see* **Note 4**).
14. Mist inside of the plastic bag with water and cover the inoculated plants.
15. Place inoculated plants in a tray with water on the bottom. Place 9–12 pots per tray.
16. Incubate in a dew chamber at 10 °C in the dark with a water bath set at 15 °C to help improve the inoculation of stripe rust for 24 h (*see* **Note 2**).
17. After, move plants to a growth chamber (or greenhouse) set at 15 °C. Leave individual bags on the plants for an additional 48 h before removing them (*see* **Note 16**).
18. Water all plants from the bottom.
19. Once urediniospores are present on the inoculated leaf, either pick one or a few well-isolated single urediniospores and repeat the process for a second cycle of purification. This way the pustules will have been purified twice and should represent a unique homogeneous genotype suitable for molecular characterization. This is critical to obtain reliable results. After the second cycle, collect the urediniospores using a mini cyclone spore collector (*see* **Note 5**), amplify the spores by reinoculating, or store them at −80 °C.

3.3 Protocol for Spore Amplification

1. Susceptible wheat plants (*see* **Note 1**) are grown in 10 cm (4″) SVD pots (20 seeds per pot) at 20 °C in greenhouse for approximately 2–3 weeks to the two-leaf stage. Plants can also be grown in the growth cabinet with a 16-h photoperiod (240 μmoles.m^{-2}.s^{-1}) at 20 °C/16 °C day/night.
2. Prior to each inoculation, clean bench and area with 70% ethanol and change gloves between rust isolates.
3. Collect fresh stripe rust urediniospores from inoculated plants. Collect rust by shaking the infected leaves inside a test tube or by vacuuming into 00 gelatin capsules or 20 mL glass vial that can be attached to a mini cyclone spore collector (*see* **Note 5**). Change gloves between rust isolates. Use approximately 10 mg spores/mL for inoculation (1 mg of spores ≈ 2.45×10^5 spores).
4. Frozen stripe rust samples do not require heat shock prior to inoculation, but we still heat shock 2 min at 45 °C (*see* **Note 17**).

5. Pour 25 mL of inoculating fluid [12, 13] into a clean 50 mL tube. Add inoculating fluid (pipet) to rust in 00 gelatin capsule or in the 20 mL glass vial that can be attached to the spore inoculator at a concentration of about 10 mg/mL (*see* **Notes 3** and **6**).
6. Mist plants with water (like dew) containing 1 drop of nonionic detergent/L prior to spraying (*see* **Note 4**).
7. Spray plants using the inoculator attached to the air compressor covering all leaves equally. Turn plants 180° and repeat spray inoculation. Set the source of compressed air or air compressor pressure to around 10–15 PSI (70–100 kP).
8. Mist plants with water (like dew) containing 1 drop of nonionic detergent/L using a water spray bottle (*see* **Note 4**).
9. Cover plants with a clear bag or tray cover for transport.
10. Move plants in a dew chamber at 10 °C in the dark for 24 h. Space pots in the dew chamber so the plants do not touch. Do not allow plants to touch the dew chamber walls. A dew chamber set at 10 °C with water bath set at 15 °C helps to improve inoculation of stripe rust (*see* **Note 2**).
11. Take plants from the dew chamber, open the end of each bag, and add stakes to help support each bag. Place plants in a tray for watering and put plants in a growth chamber at 15 °C for optimal rust infection.
12. Water all plants from the bottom. Take care to not disturb the lower leaves with the hose and risk losing spores.
13. Rust infection appears around 12 days post inoculation. Collect rust every 3 days from 15 to 28 days post inoculation (*see* **Note 5**).
14. Inoculate new plants with the freshly collected spores (*see* **Note 6**) or immediately freeze them at −80 °C.

3.4 Determination of Virulence/Avirulence Pattern of Stripe Rust Isolates

1. Use spray inoculation as above (use approximately a spatula tip of rust in ½ vial of inoculation fluid [should be like a rust-colored weak tea] [*see* **Note 3**]). This should be sufficient to spray four trays (two sets of 18 differential Avocet lines) [13, 16, 17].
2. Make sure to thoroughly clean the sprayer with ethanol between each rust sample. If limited rust, then use the fingertip inoculation method describe above to inoculate the largest leaf available.
3. The virulence/avirulence spectrum of these isolates is identified using infection results on Avocet near-isogenic lines carrying single resistance genes and some additional differential lines [13, 16, 17].

3.5 DNA Isolation from Stripe Rust Spores and DNA Sequencing

1. Breaking up spores: In 2 mL tough conical tubes (*see* **Note 8**), combine 80–100 mg of stripe rust spores with 1 g of 3 -mm-diameter synthetic beads, two larger synthetic beads (2 mm diameter), and 30–35 mg diatomaceous earth. Flash-freeze in liquid nitrogen, and then pulverize plant tissue in pre-chilled grinder (*see* **Notes** 7 and **18**). After grinding, return tubes to liquid nitrogen.
2. Any DNA extraction kit or your favorite DNA extraction protocol compatible with DNA sequencing can be used to obtain good-quality DNA (*see* **Note 11**).
3. DNA samples should have a good level of integrity as assessed by gel electrophoresis and OD 260/280 ratio of 1.8 to 2.0 and OD 260/230 ratio of at least 2.0. Routinely, DNA samples were resuspended in 10 mM Tris–HCl, pH 8.0, with 0.1 mM of EDTA at a concentration between 50 and 100 ng/μL. We routinely supplied between 150 and 500 ng/sample which represented an excess as sequencing reaction can be carried out with less. It is important that the quality and concentration of DNA and quantity are aligned with your in-house/service provider requirements. Illumina paired-end sequencing (100 bp) was carried out in a sequencing facility or in-house (*see* **Note 12**). The advantage of HiSeq sequencing over MiSeq was a higher depth of sequencing information and the possibility of multiplexing different samples to considerably reduce the cost per sample. Samples with a minimum of 16 Mb raw data provided sufficient data for in-depth analysis of individual stripe rust isolates.
4. Raw paired-end reads were trimmed and cleaned for quality and artifacts using fastx and fastq tools. For fastx the following parameters were selected: fastx_clipper; fastx_artifacts_filter; fastq_quality_trimmer; fastq_quality_filter; minimum sequence length, 20; remove adapter sequences; discard, too-short reads, adapter-only reads, N reads, or low-quality reads; quality cut-off, 20, minimum % of bases that must have 20; and quality, 50% using standard default conditions. Geneious Prime can also be used to trim and remove artifacts from reads.
5. For sequence assembly, clean reads were de novo assembled, mapped to genomic and transcriptomic reference sequences of PST-78 and Pst-CY32, and compared locally using the tool suite in Geneious (http://www.geneious.com) on both PC and server versions (*see* **Note 13**). Open-source software such as bwa, Mugsy, SignalP, InterProScan, and Blast2GO were also used (*see* **Note 14**). Consensus sequence alignments and phylogenetic trees were generated in Geneious. Random sequences were concatenated prior to sequence alignment.

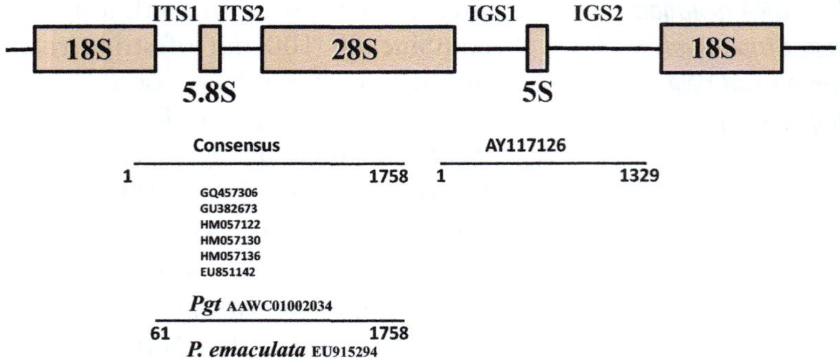

Fig. 1 Schematic representation of the rDNA gene locus in *Puccinia striiformis* for four rDNA genes—18S, 28S, 5.8S, and 5S—the internal transcribed spacers (ITS), and the intergenic spacers (IGS). The reference sequence for analysis was based on a consensus of six stripe rust sequences for the consensus region from 18S to 28S and one sequence from 28S to 18S. The *Puccinia graminis* (*Pgt*) sequence reported is from the Broad Institute and EU915294 corresponds to *Puccinia emaculata*

3.6 Case Study

1. A schematic representation of the rDNA gene locus in *Puccinia striiformis* for the four rDNA genes (18S, 28S, 5.8S, and 5S), the internal transcribed spacers (ITS), and the intergenic spacers (IGS) that were needed to establish the analysis of different stripe rust isolates in this case study is presented (Fig. 1). The reference sequence for analysis was based on a consensus of six stripe rust sequences (GQ457306, GU382673, HM057122, HM057130, HM057136, EU851142) for the consensus region from 18S to 28S and one sequence (AY117126) from 28S to 18S.
2. In this cases study, the virulence of two stripe rust isolates is presented in Table 1 based on inoculation of Avocet susceptible (AV) and 19 different Avocet near-isogenic lines containing 1 specific stripe rust resistance gene (*Yrx*) each. A phylogenetic tree using MAFFT-UPGMA was obtained using information from 16 isolates where the corresponding stem rust (*Pgt*) sequence was used as an outlier (Fig. 2). However, it is difficult to resolve closely related isolates when using rDNA sequences. To improve the resolution among the stripe rust isolates, we used a total of 1700 random sequences selected to include more than 50 nucleotides in length. Using this approach, we were able to resolve the different stripe rust isolates (Fig. 3). Our results show that sequencing is an easy and very useful tool to distinguish closely related isolates of stripe rust. This approach is directly applicable to all isolates of plant fungal pathogens.

Table 1
Virulence of two stripe rust isolates, Foremost 2011 and LSW3_2012, on 20 Avocet wheat lines

Identifier	Av	*Yr1*	*Yr5*	*Yr6*	*Yr7*	*Yr8*	*Yr9*	*Yr10*	*Yr11*	*Yr12*	*Yr15*	*Yr17*	*Yr18*	*Yr23*	*Yr24*	*Yr26*	*Yr36*	*YrSP*	*YrSK*	*YrA*
Foremost 2011	+			+	+	+		+	+	+		+	+	+			+			+
LSW3_2012	+			+	+	+	+	+	+	+		+	+	+		+	+		+	+

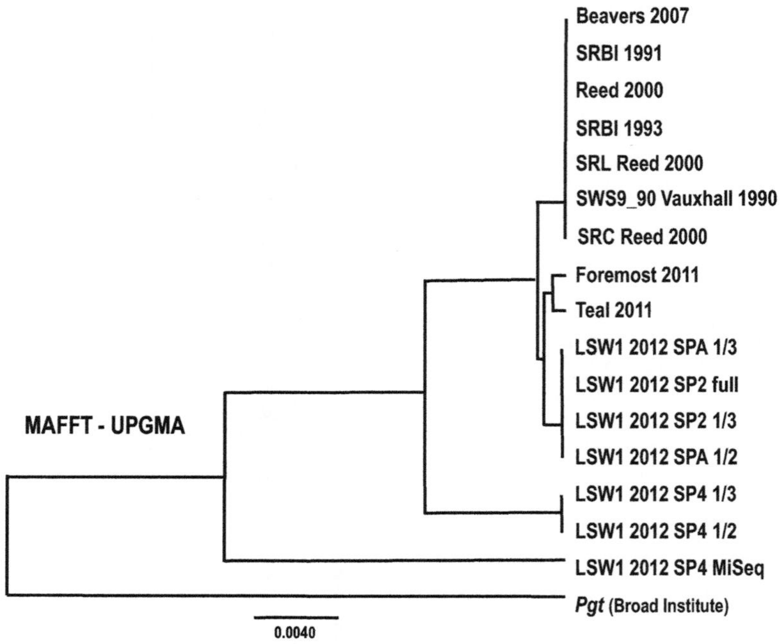

Fig. 2 Phylogenetic tree based on the consensus rDNA sequence identified in Fig. 1 covering the 18S, ITS1, 5.8S, ITS2, 28S and the 28S IGS1, 5S, IGS2 regions using MAFFT which is a multiple sequence alignment program based on fast Fourier transformation (FFT). The corresponding stem rust (*Pgt*) sequences obtained from the Broad Institute were used to generate the tree. The concatenated sequences are first subjected to an all-to-all pairwise comparison implemented in the MAFFT program. The pairwise distances are used for initial clustering by UPGMA or minimal linkage. The resulting tree is intended to roughly visualize the treelike relationship for even large numbers of sequences and to provide visual guidance for the next steps and is displayed with a Java applet version of the Archaeopteryx phylogenetic tree visualization and analysis tool. The MSA of the selected sequences is calculated by MAFFT and subjected to phylogenetic tree inference with the neighbor-joining (NJ) method

4 Notes

1. Spring wheat Avocet S is an excellent stripe rust-susceptible wheat cultivar as it is used around the world for growth chamber or greenhouse studies. For field studies, a locally adapted susceptible wheat variety is selected.
2. If a dew chamber is not available, you can incubate inoculated plants in a 10 °C cold room for 24 h under darkness. To help maintain a high level of humidity, make a "hat" with a black plastic garbage bag by rolling the end, and cover all plants in the tray.
3. You could substitute Novec 7100 Engineered Fluid (Millipore Sigma Cat # SHH0002) inoculation fluid for Soltrol®

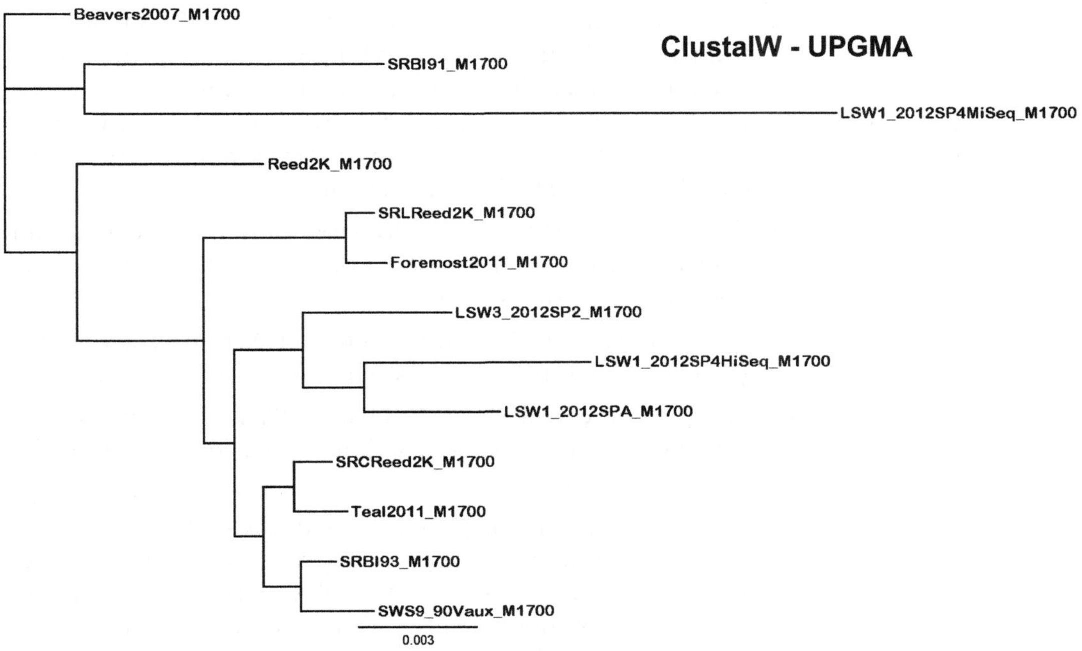

Fig. 3 Phylogenetic tree based on a total of 1700 MiSeq selected random sequences with more than 50 nucleotides mapped to the PST78 contig which were concatenated for the phylogenetic analysis (*see* **Note 19**)

170 Isoparaffin (Chevron Phillips Chemical) [12, 13]. We have observed that when using field samples (possibility of many different pathogens present), the quantity of powdery mildew (*Blumeria graminis* f. sp. *tritici*) decreases on inoculated leaves when total spores are suspended with Novec 7100 fluid rather than Soltrol® 170 Isoparaffin. These are the two universally used fluid for effective inoculation of wheat rust fungal pathogens.

4. We have routinely used Tween 20.
5. We have used a mini cyclone spore collector (# GRA-201) obtained from Tallgrass Solutions Inc., Manhattan, Kansas 66502, USA (www.tallgrassproducts.com). Recently, a similar device has been described by Pretorius et al. (2019) [18]. Alternatively, a similar device can be custom-made by a glassblower.
6. We have used a spore inoculator (# GRA-301) obtained from Tallgrass Solutions Inc., Manhattan, Kansas 66502, USA (www.tallgrassproducts.com). Recently, a similar device has been published by Pretorius et al. (2019) [18]. Alternatively, a similar device can be custom-made by a glassblower.
7. We have successfully used a Precellys 24 bead-beating grinding instrument (www.bertin-instruments.com) as stripe rust spores are very difficult to break open. We operate the grinder for two consecutive times for 10 s at 6200 rpm.

8. We have used with success the 2 mL reinforced conical tubes from Bertin Instruments cat #MK28. Other suppliers could also provide a satisfactory 2 mL conical tube. A reinforced tube is critical as otherwise, regular 2 mL conical tubes will break during the grinding process and samples will be coating the chamber of the grinding instrument.
9. We have successfully used Zirconia beads 1 mm and 2 mm (BioSpec Products cat# 11079110zx, 11079124zx) to break open the stripe rust spores.
10. We have successfully used Celite S (Diatomaceous Earth) (Sigma-Aldrich) to help break open the stripe rust spores.
11. We have successfully used the DNeasy Plant Mini Kit isolation (QIAGEN). Any protocol yielding good-quality DNA for sequencing should work.
12. We have used Illumina paired-end sequencing technology but other high-throughput sequencing technologies are available. HiSeq reads were from the McGill University and Génome Québec Innovation Centre, Montréal, using their recommended protocol. Some isolates were sequenced in-house using a MiSeq and the Nextera DNA sequencing protocols. Alternative sequencing technologies will also provide reliable information.
13. We have successfully used Geneious or Geneious Prime software (http://www.geneious.com) either in a PC or server version to carry out most analyses. Other packages from different suppliers can be also used satisfactorily.
14. Open-source software such as bwa, Mugsy, SignalP, InterProScan, and Blast2GO were also used under the standard setting conditions. Other equivalent open-source packages could be also used satisfactorily.
15. Only a few minutes if using Novec 7100 fluid [12].
16. When doing multiple purification of pustules, we leave the bag on each pot to minimize potential cross-contamination.
17. Frozen leaf rust does require a heat shock (2 min at 45 °C) prior to inoculation, so by skipping this step, this can be used to eliminate leaf rust contamination of field samples.
18. Stripe rust spores are very difficult to break open. This is the protocol that offers the highest yield of extracted DNA. The DNA is long enough for PacBio sequencing analyses when isolated with a phenol-chloroform extraction.
19. It is important to carefully verify the concatenation by pairwise comparisons before launching the analysis.

References

1. Dean R, Van Kan JL, Pretorius ZA et al (2012) The top 10 fungal pathogens in molecular plant pathology. Mol Plant Pathol 13:414–430
2. McIntosh RA, Wellings CR, Park RF (1995) Wheat rusts: an atlas of resistance genes. CSIRO Publications, East Melbourne
3. Peterson PD (2013) "The barberry or bread": the public campaign to eradicate common barberry in the United States in the early 20th century. APS Features https://doi.org/10.1094/APSFeatures-2013-08. Accessed 17 March 2023
4. McCallum BD, Hiebert CW, Cloutier S et al (2016) A review of wheat leaf rust research and the development of resistant cultivars in Canada. Can J Plant Pathol 38:1–18
5. Lei Y, Wang M, Wan A et al (2017) Virulence and molecular characterization of experimental isolates of the stripe rust pathogen (*Puccinia striiformis*) indicate somatic recombination. Phytopathology 107:329–344
6. Jin Y, Szabo LJ, Carson M (2010) Century-old mystery *of Puccinia striiformis* life history solved with the identification of *Berberis* as an alternate host. Phytopathology 100:432–435
7. Milus EA, Seyran E, McNew R (2006) Aggressiveness of *Puccinia striiformis* f. sp. *tritici* isolates in the south-central United States. Plant Dis 90:847–852
8. Chen XM (2005) Epidemiology and control of stripe rust [*Puccinia striiformis* f. sp. *tritici*] on wheat. Can J Plant Pathol 27:314–337
9. Hovmøller MS, Walter S, Bayles RA et al (2016) Replacement of the European wheat yellow rust population by new races from the centre of diversity in the near-Himalayan region. Plant Pathol 65:402–411
10. Ali S, Rodriguez-Algaba J, Thach T et al (2017) Yellow rust epidemics worldwide were caused by pathogen races from divergent genetic lineages. Front Plant Sci 8:1057
11. Khanfri S, Boulif M, Lahlali R (2018) Yellow rust (*Puccinia striiformis*): a serious threat to wheat production worldwide. Not Sci Biol 10: 410–423
12. Sørensen CK, Thach T, Hovmøller MS (2016) Evaluation of spray and point inoculation methods for the phenotyping of *Puccinia striiformis* on wheat. Plant Dis 100:1064–1070
13. Mallard S, Gaudet D, Aldeia A et al (2005) Genetic analysis of durable resistance to yellow rust in bread wheat. Theor Appl Genet 110: 1401–1409
14. Laroche A, Frick M, Puchalski B et al (2015) Characterization of stripe rust isolates from western Canada using NGS. Paper presented at the 23rd international plant and animal genome conference, Town and Country Resort and Conference Center, San Diego, 10–14 January 2015
15. Hubbard A, Lewis CM, Yoshida K et al (2015) Field pathogenomics reveals the emergence of a diverse wheat yellow rust population. Genome Biol 16:23
16. Washington State University (2022) Stripe rust lab, Pullman. https://striperust.wsu.edu/races/differential-cultivars-of-wheat/. Accessed 17 March 2023
17. Su H, Conner RL, Graf RJ, Kuzyk AD (2003) Virulence of *Puccinia striiformis* f.sp. *tritici*, cause of stripe rust on wheat, in Western Canada from 1984–2002. Can J Plant Pathol 25:312–319
18. Pretorius ZA, Booysen GJ, Boshofft WHP et al (2019) Additive manufacturing of devices used for collection and application of cereal rust urediniospores. Front Plant Sci 10:639

Chapter 3

DNA-Barcoding Identification of Plant Pathogens for Disease Diagnostics

Nicolas Feau, Padmini Herath, and Richard C. Hamelin

Abstract

The accurate identification of plant pathogens is a critical step to prevent their spread and attenuate their impact. Among the wide range of methods available, DNA-barcoding, i.e., the identification of an organism through the PCR amplification and sequencing of a single locus, remains one of the most straightforward and accurate plant-pathogen identification techniques that can be used in a generic molecular biology lab. This chapter provides a detailed protocol for the isolation of genomic DNA of fungal and oomycete pathogens from fresh field samples and the amplification and sequencing of the internal transcribed spacer (ITS) locus for DNA-barcoding purpose. Amendments to the protocol are provided to help in resolving issues related to the analysis of complicated samples and to the lack of species resolution that can be encountered with ITS barcodes.

Key words DNA-barcode, Molecular identification, ITS, Translation elongation factor 1-alpha 1 (EF-1α), Fungal pathogen, Oomycetes, *Phytophthora*, *Pythium*, Bioinformatics

1 Introduction

Early pest detection and identification is a cornerstone for prevention of disease outbreaks. This is particularly relevant as the risk posed by emerging plant diseases and pest outbreaks is increased by anthropogenic factors such as global trade and climate change [1–3]. However, accurate plant pathogen identification is challenging. The taxonomic diversity of plant pathogens that threaten crops and trees is breathtaking and morphology-based identification of these organisms requires expertise that is difficult to learn and maintain across such a broad taxonomic range. Plant pathogens are broadly organized in four main taxonomic groups (fungi, oomycetes, bacteria, and virus), among which the fungi and oomycetes are responsible for 70% to 85% of plant diseases [4]. Among these, some specific groups of pathogens with an obligate biotrophic lifestyle, such as the fungal rusts (Pucciniales) and powdery mildews

Nora A. Foroud and Jonathan A. D. Neilson (eds.), *Plant-Pathogen Interactions*, Methods in Molecular Biology, vol. 2659, https://doi.org/10.1007/978-1-0716-3159-1_3,

(Erysiphales), cannot be maintained in culture on artificial media, making their morphological identification even more complicated. In addition, many plant pathogens are part of complexes of "cryptic species" that are difficult to distinguish on a morphometrical basis, but are differentiated based on molecular traits; however, these closely related species may harbor different biological traits and lifestyles. For example, the fungal pathogen *Hymenoscyphus pseudoalbidus*, which is on the list of regulated pests in Canada and the USA, cannot be visually or morphologically differentiated from the nonpathogenic sister species *H. albidus* [5, 6]. In quarantine procedures, the confusion between these two species would have major economic, social, and/or environmental consequences, and an accurate identification method (e.g., molecular marker) is essential to recognize them.

Innovations in molecular biology and genomics have provided tools that can address plant pathogen identification and detection challenges. Polymerase chain reaction (PCR) tests provides multiple advantages, including (1) less knowledge and expertise in mycology required than in classical culture-dependent approaches, (2) application to all taxonomic groups of plant pathogens, (3) not dependent on obtaining pure cultures, and (4) can be conducted by a broad range of users equipped with a generic molecular biology lab. In the context of plant pathogen diagnostics, PCR-related methods can be used in two ways: for identification when the pathogen is unknown and for detection of a known or targeted pathogen. The DNA-barcoding method uses a short, standardized DNA fragment providing a high, interspecific variability and low intraspecificity that enable the identification of organisms at the species level [7]. The identification is achieved by matching the DNA barcode with species-referenced barcodes in DNA sequence databases such NCBI nonredundant (nr). Over the last decade, application of DNA barcoding has been tested and standardized for plant quarantine pathogens [8, 9] and has allowed detection of symptomless and latent plant pathogens [10] and has enabled identification of major crop and tree pathogens [11–13]. In particular, the association of DNA barcodes with reference specimen (i.e., type, ex-type, and voucher) from fungal and plant pathogen herbaria as well as with curated fungal sequence databases [e.g., [14] has allowed researchers to revisit the taxonomy and identification of specimen of phytosanitary concern [15, 16].

To be practical as a DNA barcode, a gene region must satisfy the following criteria: (1) must contain a "barcoding gap," i.e., substantial genetic divergence between species and minimal variability within, (2) must have conserved flanking sites to develop universal PCR primers for the widest taxonomic application, and (3) must be of appropriate sequence length to be accurately amplified by PCR and sequenced. Ideally, the selected marker should be present in several copies per cell in the targeted organism, facilitating detection by allowing a high sensitivity of the PCR amplification. The internal transcribed spacer (ITS) that lays between the conserved flanking regions of the small and large subunit of ribosomal RNA is the most powerful and standard barcode in eukaryotes that has been officially proposed as a core DNA barcode for fungi [17]. However, it has some limitations. In some taxonomic groups, the ITS either is too polymorphic within species or does not contain enough variation to provide resolution among species [18–20]. Specifically, for fungal and oomycete genera rich in complexes of cryptic species (e.g., *Fusarium*, *Ceratocystis*, and *Ophiostoma* in the ascomycetes, *Armillaria* and *Heterobasidion* in the basidiomycetes), there is insufficient variation in the ITS cluster to accurately identify species [21–24]. For these genera, a second barcoding gene such as one of those traditionally used to build species phylogenies [e.g., *β-tubulin*, *EF-1α*, and *RPBII*; 25] is required for precise species identification.

Even in the era of whole genome sequencing, DNA barcoding still provides a rapid, cheap, specific, accurate, and analytically simple method of plant pathogen identification. Here, we provide a mainstream protocol for isolation of genomic DNA from plant tissues infected by fungal or oomycete pathogens and its use in polymerase chain reaction (PCR)-based amplification of the ITS region between the conserved flanking regions of the small and large subunit of ribosomal RNA for barcoding purposes (Fig. 1). The DNA-extraction protocol is amended with alternative procedures, to allow purification of cleaner and non-contaminated DNA. Additional primers and protocol for the use of the EF-1α as a secondary barcode [26] is proposed for plant pathogen genera for which the interspecific divergence within the ITS locus proved to be insufficient.

2 Materials

For all the procedures described below, the user should wear personal protection equipment (e.g., lab coat and disposable gloves). Use of separate rooms for sample processing, culturing, and pre- and post-PCR steps is strongly recommended to avoid contamination by exposing the samples to an environment where PCR products were handled. Use sterile pipettes and pipette tips that are only used for DNA extraction and separate pipette sets for DNA-extraction, PCR, and electrophoresis steps. Additionally, sequence editing software of the user's preference will be required (*see* **Note 1**).

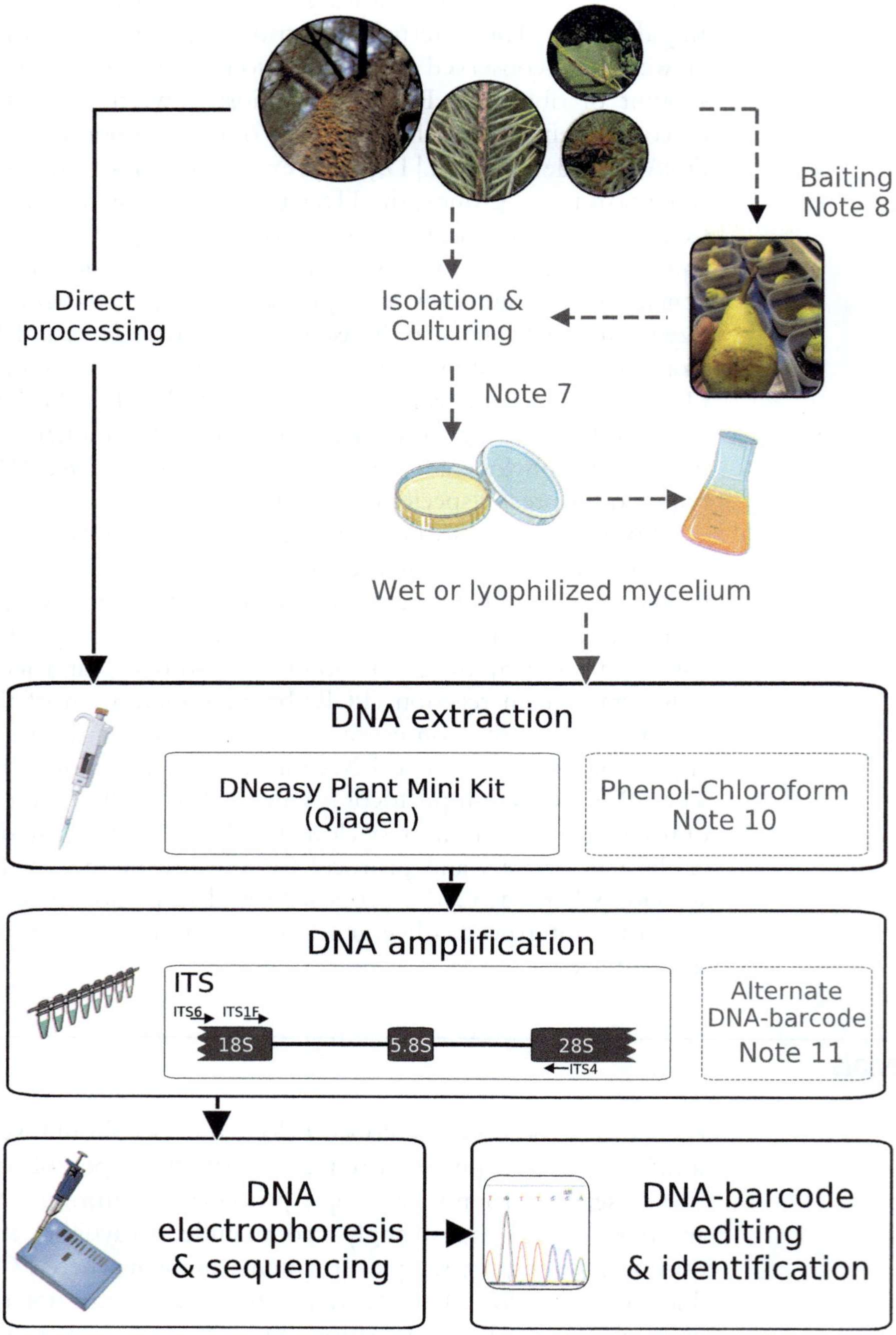

Fig. 1 DNA barcoding of plant pathogens. Mainstream process (i.e., direct DNA extraction and PCR) of infected plant tissues is indicated with black text and arrows. Alternative processes referring to **Notes 7**, **8**, **10**, and **11** are indicated in light grey

2.1 Sample Preparation and DNA Extraction

1. Dissecting microscope.
2. Sterile scalpels, tweezers, and inoculating needles.
3. Clean and sterile 2 mL safe-lock and 1.5 mL microcentrifuge tubes.
4. Pipettes (10–1000 μL) with disposable pipette tips.
5. 3.2 mm-diameter stainless steel beads.
6. Tissue homogenizer and tissue homogenizer adapters (stored at −20 °C in a freezer until utilization) (*see* **Note 2**).
7. Digital block heaters or water bath.
8. Microcentrifuge (~20,000× *g*).
9. Vortex.

2.2 DNA Concentration and Quality Control

1. 10 μL pipette with disposable pipette tips.
2. Spectrophotometer to measure absorbance ratio A260/280 nm and A260/A230 nm (*see* **Note 3**).
3. Sterile distilled water.
4. Vortex.

2.3 PCR Amplification

1. Pipettes (10–1000 μL) with disposable barrier tips.
2. Standard thermocycler.
3. PCR plasticware, either microtubes strips with caps or 96-well microplates.
4. Vortex.
5. Mini table centrifuge.
6. DNA polymerase (*see* **Note 4**).
7. Primers.

2.4 Agarose Gel Electrophoresis

1. 1× TBE (89 mM boric acid; 2 mM EDTA (pH 8), 89 mM Tris base).
2. Ultrapure agarose.
3. Gel casting tray.
4. 100-bp DNA ladder.
5. Loading dye (*see* **Note 5**).
6. UV or blue light illuminator (*see* **Note 5**).

3 Methods

3.1 Sample Preparation and DNA Extraction

1. Fresh samples should be prepared for DNA extraction as follows, depending on the type of plant material.
2. For necrotic roots: Roots should be carefully washed under tap water, rinsed with sterile distilled water and dried briefly on paper towels. Cut two to five 5 mm-long pieces depending on the diameter of the roots.
3. For leaves with surface necrotic lesions or leaf spots: Leaves should be rinsed with sterile distilled water and dried briefly on paper towels. Cut off disks of 5 mm diameter at the margin of the lesion using a cork borer.
4. For internal lesions (necrotic cankers, infected wood samples): Rinse the material (stem and branch pieces) with sterile distilled water, and excise (in sterile condition) a piece of five-by-five millimeters of tissue under the surface at the edge of the lesion.
5. For rust pustules on leaves: Pustules should be directly excised without being rinsed. For those sporulating, the spore content can be harvested with clean Q-tips and transferred into a 2 mL tube.
6. For sporulating samples on the surface of a lesion (either on branches, stems, or leaves): Spores can be directly transferred with a sterile fine-gauge inoculating needle for DNA extraction or to a growth medium (*see* **Notes 6–8**).
7. Transfer collected sample(s) into a tissue homogenizer of choice and homogenize the samples according to manufacturer's protocol (*see* **Note 9**).
8. Extract the genomic DNA (gDNA) with user-preferred protocol (*see* **Note 10**). Ensure DNA is eluted in sterile distilled water rather than buffer.

3.2 Measurement of DNA Concentration and Quality Control

1. Assess DNA quality and concentration using a spectrophotometer (*see* **Note 3**). Samples that do not meet QC itemized below should be rejected. DNA yield and quality of those samples can be improved by using alternate grinding and DNA-extraction protocols (*see* **Notes 9** and **10**).
2. DNA concentration in ng/μL can range between 2 ng/μL and 100 ng/μL, depending on the type of tissue sampled.
3. Purity of the DNA from proteins is indicated by the absorbance ratio A260/A280 nm; this value should fall between 1.7 and 1.9.
4. Purity of the DNA from carbohydrates (and eventually phenol and EDTA) is indicated by the absorbance ratio A260/A230 nm and should fall between 2.0 and 2.3.

3.3 PCR Amplification of the ITS Barcode Region

1. Primer sequences [27, 28] are presented in Table 1. Additional primers for an alternate barcode are presented (*see* **Note 11**).
2. PCR reagents and reaction volumes optimized for Platinum *Taq* DNA polymerase (*see* **Note 4**) are presented in Table 2. Other DNA polymerases can also be used according to the manufacturer's specifications.
3. PCRs are performed in a standard thermal cycler with an initial denaturation step of 94 °C for 3 min, followed by 35 cycles of denaturation (30 s at 94 °C), annealing (30 s at 55 °C), extension (1 min at 72 °C), and a final extension at 72 °C for 10 min. Other reaction conditions may be required for different DNA polymerases.

Table 1
Universal primers targeting the internal transcribed spacers (ITSs) for DNA barcoding of fungal and oomycete pathogens

Target	Primer	Primer sequence 5′->3′	References
Fungi			
	ITS1F (forward)	CTTGGTCATTTAGAGGAAGTAA	[27]
	ITS4 (reverse)	TCCTCCGCTTATTGATATGC	[28]
Oomycetes			
	ITS6 (forward)	GAAGGTGAAGTCGTAACAAGG	[29]
	ITS4 (reverse)	TCCTCCGCTTATTGATATGC	[28]

Table 2
Reagent volumes and concentration for PCR amplification using the platinum *Taq* DNA polymerase

Reagent[a]	Volume per reaction
10× buffer, no $MgCl_2$ (Invitrogen, Carlsbad, CA, USA)	2.5 μL (1×)
$MgCl_2$ (50 mM) (Invitrogen)	0.75 μL (1.5 mM)
Forward primer (10 μm)	2.5 μL (1 μM)
Reverse primer (10 μm)	2.5 μL (1 μM)
dNTP mix (10 mM) (Invitrogen)	0.5 μL (0.2 mM of each dNTP)
Platinum *Taq* DNA polymerase (5 U/μL) (Invitrogen)	0.1 μL (0.02 U/mL)
DNA sample (2–10 ng/μL)	2 μL
Sterile milliq water	14.15 μL
Total volume	25 μL

[a]The regent volumes and concentrations described in this table may vary according to the polymerase of choice

3.4 Assessment of PCR Product Size and Quality by Agarose Gel Electrophoresis

1. Prepare a 1.5% agarose gel by dissolving 0.6 g agarose powder, in 40 mL of 1× TBE in a standard casting tray (*see* **Note 12**).
2. Load 5 μL of the PCR product (1:5, loading dye/PCR product) onto the gel. Use a 100-bp ladder for DNA fragment size reference.
3. Run the gel at 100 V for 30–45 min and visualize the gel under blue or UV light to determine gDNA size and quality of the PCR product. A unique, sharp, and bright band indicates a high-quality product.
4. A single band is expected. If there are multiple bands, or no bands, then the PCR conditions provided in Subheading 3 should be modified, by increasing the concentration of $MgCl_2$- and/or decreasing the annealing temperature to reduce primer specificity to the DNA target (however, this increases the risk of amplifying nontarget DNA regions). For fungi, the expected band size should be around 500 to 750 bp [29]; Phytophthoras and Pythiums have usually longer products, varying from 750 to 815 bp with ITS4/ITS6 [30].
5. An aliquot of the PCR product can be sent for cleaning and Sanger sequencing through a sequencing facility. For each PCR product, we recommend sequencing in both forward (ITS1F) and reverse direction (ITS4) to facilitate sequence editing and increase taxon assignment accuracy.

3.5 Sequence Editing and Taxon Assignment to the DNA Barcode

1. Raw sequencing chromatograph files (extension.*ab1*) can be edited into text format (FASTA) using BioEdit 7.2 software [31]. Sequence editing is described in **steps 2–6** (*see* **Note 1**).
2. Check the quality of each chromatograph file (*Import* > *Sequence* in BioEdit; the chromatograph file will automatically open and appear with a sequence file). Troubleshoot (and eventually discard) chromatographs with weak sequencing signal and/or strong baseline noise (*see* **Note 13**).
3. Reverse-complement the trace/sequence obtained with the reverse primer (ITS4) (*Sequence* > *Nucleic acid* > *Reverse complement* in BioEdit).
4. Align the forward sequence (ITS1F) with the reverse-transcribed sequence.
5. Trim the 5′- and 3′-ends of the obtained sequence reads, based on visual inspection (make sure Bioedit is in *Edit* mode).
6. Search for (a) heterozygous (or "double") peaks at single positions (common when sequencing a DNA barcode from a diploid or dikaryotic organisms [oomycetes and basidiomycetes]) and correct with the appropriate IUPAC nucleotide code and (b) sequence conflicts between the forward and

reverse-complement sequence and correct based on visual inspection of the chromatograph files (*see* **Note 13** about heterozygote indels).

7. Search the edited sequence against the NCBI nr database for rapid taxonomic identification using BLASTn with default parameters.
8. Check the first 20–50 hits of the BLAST results to assess the reliability of a GenBank match. Ideally, type, ex-type, and voucher specimen should be present among these hits; published sequences should also be present and are usually more reliable than unpublished ones.
9. For a better taxonomic assignment of the DNA-barcode, assemble an alignment with the 20–50 top sequences from the BLASTn result and construct a phylogenetic tree. This should help to identify the smallest statistically supported group of sequence, including the DNA-barcode sequence, whose members are descended from a common ancestor (i.e., phylogenetic species concept; [32]) (*see* **Note 14**).

4 Notes

1. The sequence editing protocol described in Subheading 3.5 is based on BioEdit 7.2.
2. Retsch® MM 300 TissueLyser or another equivalent instrument can be used with 2 mL tubes containing one 3.2 - mm-diameter stainless steel bead.
3. NanoDrop™ ND 1000 Spectrophotometer (Thermo Fisher) or another equivalent instrument can be used.
4. Low-error rate *Taq* polymerases such as Platinum™ Taq DNA Polymerase High Fidelity are preferable to reduce errors during DNA elongation.
5. SafeGreen™ dye can be use with Blue Light (e.g., PrepOne Sapphire Blue LED Illuminator system [Embi Tec, CA, USA]) or UV transilluminator to avoid using ethidium bromide.
6. Surface sporulation from lesions on the plant tissues can be induced by incubating the sample in a moist chamber (can be as simple as placing the plant tissue on sterile filter paper moistened with distilled water and placing it in a sterile Petri dish sealed with parafilm) at room temperature for 24 h. DNA can be directly extracted from masses of spores exuded from fruiting bodies that appear on the plant tissues. Spores can also be used for fungal isolation (*see* **Note** 7).
7. Fungal pathogens other than obligate biotrophs (i.e., powdery mildews and fungal rusts) can be isolated in pure cultures. This

can be achieved by surface cleaning excised pieces of infected tissue (e.g., 2 min in 70% ethanol or 0.5% bleach, rinse with sterile distilled water) and placing it on suitable agar growth medium in sterile Petri dishes. Many isolation protocols and selective media have been described for plant pathogens [33, 34]. A pure culture can then be transferred into liquid media to produce fungal biomass and facilitate DNA extraction. For *Phytophthora* spp. and *Pythium* spp., pure cultures can be obtained using a "baiting" technique (*see* **Note 8**).

8. *Phytophthora* and *Pythium* plant pathogens of the oomycete group can be baited from symptomatic material, soil, and water by using rhododendron leaves or pear baits [35, 36] and plating lesions on baits with an oomycete-specific agar growth medium such as CMA-PARP(H) [37].
9. For difficult samples, grinding the mycelia with a mortar and pestle under liquid nitrogen should result in higher yields of DNA.
10. QIAGEN's DNeasy® Plant Mini kit and DNeasy® Plant Pro Mini kit (Qiagen Inc., Toronto, ON, Canada) can be used for high-throughput DNA extractions by following the manufacturer protocol for purification of total DNA from fresh plant tissue (Mini protocol) (Qiagen DNeasy® Plant Handbook, 07/2020, p:27). Use a volume of 50 μL sterile distilled water as the final elution solvent instead of buffer AE. For higher yields, a phenol-chloroform DNA extraction protocol [38] may be used.
11. The EF-1α has been proposed as an alternate DNA barcode for fungal genera with species that cannot be differentiated with the ITS locus [26]. Table 3 includes universal EF-1α- primers [26, 39, 40] that can be used with the PCR protocol described in the *PCR Amplification of the ITS Barcode Region* section.
12. Manipulation of PCR products such as gel electrophoresis should be carried out in a separate room from the sample processing and PCR rooms.
13. Sequencing chromatographs with "overlapping peaks" and/or strong baseline "noise" can be results from different issues: (a) Overall sequencing signal is weak, resulting in strong background noise. The sequence should be rejected. This can be caused by the presence of a chemical contaminant(s) in the DNA sample; this issue can be fixed by using a CTAB/phenol-chloroform DNA-extraction protocol (*see* **Note 10**). (b) Loss of sequencing signal along the sequence: the

Table 3
Universal primers for the PCR amplification and sequencing of the EF-1α-barcode in fungi and/or ascomycetes

Primer sequences (forward/reverse)[a]	Modifications to the PCR protocol described in the *PCR Amplification of the ITS barcode region* section	References
EF1-1018F GAYTTCATCAAGAA CATGAT/EF1-1620R GACGTT GAADCCRACRTTGTC	Annealing temperature: 58 °C	[26]
EF1-983F GCYCCYGGHCAYCGT GAYTTYAT/EF1-1567R ACHGT RCCRATACCACCRATCTT	–	[39]
EF1-728F CATCGAGAAGTTCGAG AAGG/EF1-986R TACTTGAAGGA ACCCTTACC	Annealing temperature: 58 °C	[40]

[a]IUPAC code for degenerate nucleotides: Y, C/T; R, A/G; H, A/C/T D; A/G/T

beginning of the sequence is resolved, but this resolution rapidly diminishes along the sequence and the chromatograph becomes uninterpretable. For the same reason as above in "a," try an alternate DNA-extraction protocol (*see* **Note 10**). (c) Overall sequencing signal is strong, but two or more sequences seem to overlap and are irregularly spaced, starting from the beginning of the sequence. Two (or more) unrelated DNA samples were sequenced together (i.e., different microorganisms in the same sample). This can be avoided by doing the PCR/sequencing on DNA isolated from a pure culture (*see* **Notes 7** and **8**). Alternatively, PCR amplicons belonging to different organisms can be cloned and sequenced individually. (d) Overall sequencing signal is strong. The first part of the sequence has no overlapping peaks, while the second part has superimposed peaks. This results from the sequencing of two related PCR amplicons (i.e., alleles) in a diploid or dikaryotic microorganism with a heterozygous indel. The two alleles can be separated using Poly peak parser if a reference sequence is available [41]. Alternatively, both PCR amplicons can be cloned and sequenced individually.

14. A rapid and user-friendly way to construct a robust phylogenetic tree is to use the NGPhylogeny server [42].

Acknowledgements

Nicolas Feau would like to acknowledge Drs. Jonathan Neilson and Nora Foroud (Agriculture and Agri-Food Canada, Lethbridge, AB, Canada) for the invitation to write this chapter. Sources of funding for this work include Genome Canada's Large-Scale Applied Research Project (LSARP project # 10106) and Biosurveillance of Alien Forest Enemies (bioSAFE; https://www.biosafegenomics.com/) awarded to Richard C. Hamelin and funding from the 2 billion trees initiative (Natural Resources Canada, Canadian Forest Service) awarded to Nicolas Feau.

References

1. Pautasso M, Schlegel M, Holdenrieder O (2015) Forest health in a changing world. Microb Ecol 69:826–842
2. Jeger M, Beresford R, Bock C et al (2021) Global challenges facing plant pathology: multidisciplinary approaches to meet the food security and environmental challenges in the mid-twenty-first century. CABI Agric Biosci 2:20
3. Hamelin RC, Roe AD (2019) Genomic biosurveillance of forest invasive alien enemies: a story written in code. Evol Appl 13:95–115
4. Carris LM, Little CR, Stiles CM (2012) Introduction to fungi. The Plant Health Instructor. https://doi.org/10.1094/PHI-I-2012-0426-0
5. Queloz V, Grünig CR, Berndt R et al (2011) Cryptic speciation in *Hymenoscyphus albidus*. For Pathol 41:133–142
6. Husson C, Scala B, Caël O et al (2011) *Chalara fraxinea* is an invasive pathogen in France. Eur J Plant Pathol 130:311–324
7. Hebert PDN, Cywinska A, Ball SL, deWaard JR (2003) Biological identifications through DNA barcodes. Proc Biol Sci 270:313–321
8. Gao R, Zhang G (2013) Potential of DNA barcoding for detecting quarantine fungi. Phytopathology 103:1103–1107
9. Bonants PJM (2014) Results of the EU project QBOL, focusing on DNA barcoding of quarantine organisms, added to an international database (Q-Bank) on identification of plant quarantine pathogens and relatives. In: Gullino M, Bonants P (eds) Detection and diagnostics of plant pathogens. Plant pathology in the 21st century (Contributions to the 9th International Congress), vol 5. Springer, Dordrecht
10. Hsieh CW, Chuang YY, Lee MZ et al (2020) First inventory of fungi in symptomless and symptomatic Chinese *Mesona* indicates phytopathological threat. Plant Dis 104:2391–2397
11. Robideau GP, De Cock AWAM, Coffey MD et al (2011) DNA barcoding of oomycetes with cytochrome c oxidase subunit I and internal transcribed spacer. Mol Ecol Res 11:1002–1011
12. Feau N, Decourcelle T, Husson C et al (2011) Finding single copy genes out of sequenced genomes for multilocus phylogenetics in non-model fungi. PLoS One 6:e18803
13. Feau N, Ramsfield TD, Myrholm CL et al (2021) DNA-barcoding identification of *Dothistroma septosporum* on *Pinus contorta* var. *latifolia*, *P. banksiana* and their hybrid in northern Alberta, Canada. Can J Plant Pathol 43:472–479
14. Schoch CL, Robbertse B, Robert V et al (2014) Finding needles in haystacks: linking scientific names, reference specimens and molecular data for fungi. Database 2014: bau061
15. Feau N, Vialle A, Allaire M et al (2009) Fungal pathogen (mis-) identifications: a case study with DNA barcodes on *Melampsora* rusts of aspen and white poplar. Mycol Res 113:713–724
16. Ikeda S, Hosoya T (2021) Re-identification of pathogen of bulb black rot of lilies based on DNA barcoding and infection studies. J Gen Plant Pathol 87:138–147
17. Schoch CL, Seifert KA, Huhndorf S et al (2012) Nuclear ribosomal internal transcribed spacer (ITS) region as a universal DNA barcode marker for fungi. Proc Natl Acad Sci 109: 6241–6246
18. Vialle A, Feau N, Allaire M et al (2009) Evaluation of mitochondrial genes as DNA barcode for Basidiomycota. Mol Ecol Res 9:99–113

19. Xu J (2016) Fungal DNA barcoding. Genome 59:913–932
20. Kiss L (2012) Limits of nuclear ribosomal DNA internal transcribed spacer (ITS) sequences as species barcodes for Fungi. Proc Natl Acad Sci 109:E1811–E1811
21. Roe AD, Rice AV, Bromilow SE et al (2010) Multilocus species identification and fungal DNA barcoding: insights from blue stain fungal symbionts of the mountain pine beetle. Mol Ecol Resour 10:946–959
22. Sharma R, Polkade AV, Shouche YS (2015) Species concept' in microbial taxonomy and systematic. Curr Sci 108:1804–1814
23. Tsykun T, Rigling D, Prospero S (2013) A new multilocus approach for a reliable DNA-based identification of *Armillaria* species. Mycologia 105:1059–1076
24. Balasundaram SV, Engh IB, Skrede I et al (2015) How many DNA markers are needed to reveal cryptic fungal species? Fungal Biol 119:940–945
25. Feau N, Vialle A, Allaire M et al (2011) DNA barcoding in the rust genus *Chrysomyxa* and its implications for the phylogeny of the genus. Mycologia 103:1250–1266
26. Stielow JB, Lévesque CA, Seifert KA et al (2015) One fungus, which genes? Development and assessment of universal primers for potential secondary fungal DNA barcodes. Persoonia Mol Phylogeny Evol Fungi 35:242
27. Gardes M, Bruns TD (1993) ITS primers with enhanced specificity for basidiomycetes – application to the identification of mycorrhizae and rusts. Mol Ecol 2:113–118
28. White TJ, Bruns T, Lee S et al (1990) Amplification and direct sequencing of fungal ribosomal RNA genes for phylogenetics. In: Innis MA, Gelfand DH, Sninsky JJ, White TJ (eds) PCR protocols: a guide to methods and applications. Academic, New York, pp 315–322
29. Manter DK, Vivanco JM (2007) Use of the ITS primers, ITS1F and ITS4, to characterize fungal abundance and diversity in mixed-template samples by qPCR and length heterogeneity analysis. J Microbiol Method 71:7–14
30. Cooke DEL, Drenth A, Duncan JM et al (2000) A molecular phylogeny of *Phytophthora* and related oomycetes. Fungal Genet Biol 30: 17–32
31. Hall TA (1999) BioEdit: a user-friendly biological sequence alignment editor and analysis program for Windows 95/98/NT. Nucleic Acids Symp Ser 41:95–98
32. Taylor JW, Jacobson DJ, Kroken S et al (2000) Phylogenetic species recognition and species concepts in fungi. Fungal Genet Biol 31:21–32
33. Tsao PH (1970) Selective media for isolation of pathogenic fungi. Annu Rev Phytopathol 8: 157–186
34. Sinclair JB, Dhingra OD (2019) Basic plant pathology methods, 2nd edn. CRC Press
35. Ivors K (2015a) PROTOCOL 01-11.1: isolation and detection of *Phytophthora* spp. using Rhododendron leaf baits. In: Ivors KL (ed) Laboratory protocols for *Phytophthora* species. APS Press, St. Paul
36. Hüberli D (2015) PROTOCOL 01-13.1: pear baiting for isolation of *Phytophthora* spp. from various substrates. In: Ivors KL (ed) Laboratory protocols for *Phytophthora* species. APS Press, St. Paul
37. Ivors K (2015b) PROTOCOL 07-03.1: PARP (H)-CMA. In: Ivors KL (ed) Laboratory protocols for *Phytophthora* species. APS Press, St. Paul
38. Hu Y (2016) High quality DNA extraction from Fungi_small scale. protocols.io. Available via https://www.protocols.io/view/high-quality-dna-extraction-from-fungi- small-scale-exmbfk6
39. Rehner SA, Buckley E (2005) A *Beauveria* phylogeny inferred from nuclear ITS and EF1-alpha sequences: evidence for cryptic diversification and links to *Cordyceps* teleomorphs. Mycologia 97:84–98
40. Carbone I, Kohn LM (1999) A method for designing primer sets for speciation studies in filamentous ascomycetes. Mycologia 91:553–556
41. Hill JT, Demarest BL, Bisgrove BW et al (2014) Poly peak parser: method and software for identification of unknown indels using sanger sequencing of polymerase chain reaction products. Dev Dyn 243:1632–1636
42. Lemoine F, Correia D, Lefort V et al (2019) NGPhylogeny.fr: new generation phylogenetic services for non-specialists. Nucleic Acids Res 47:W260–W265

Chapter 4

Real-Time Portable LAMP Assay for a Rapid Detection of *Xylella fastidiosa* In-Field

Nicola Luchi, Duccio Migliorini, Francesco Pecori, and Alberto Santini

Abstract

Early diagnosis is part of a decision-making process which in the case of plant diseases may prevent the spread of invasive plant pathogens and assist in their eradication. Significant advantages could be obtained from moving testing technology closer to the sampling site, thereby reducing the detection time. This chapter describes a portable real-time LAMP assay for a specific detection of *Xylella fastidiosa* in-field. The LAMP assay, including DNA extraction, allows a complete and specific in-field analysis in just 40 minutes, enabling the detection of pathogen DNA in host tissues.

Key words LAMP, Invasive pathogen, Detection, Isothermal amplification, Olive quick decline syndrome, Bacterial pathogens, Pierce's disease

1 Introduction

An effective framework for early warning and rapid response is a crucial element to prevent or mitigate the impact of biological invasions of plant pathogens, especially at ports of entry. Molecular detection of pathogens by using PCR-based methods usually requires a well-equipped laboratory. Rapid detection tools that can be applied as point-of-care diagnostics are highly desirable, especially to intercept quarantine plant pathogens.

Xylella fastidiosa is a gram-negative bacterium associated with serious diseases in a wide range of hosts in the world [1–4]. The disease was considered to be confined in America until 2013 when it was detected for the first time in Europe (Southern Italy–Apulia) causing the olive quick decline syndrome (OQDS) with the loss of a thousand olive trees in Apulia, with a significant socioeconomic impact [5, 6].

Since then, between 2015 and 2017, several subspecies and strain of *X. fastidiosa* were also reported in other EU countries (France, Spain, and Germany) causing damages on different host

Nora A. Foroud and Jonathan A. D. Neilson (eds.), *Plant-Pathogen Interactions*, Methods in Molecular Biology, vol. 2659,
https://doi.org/10.1007/978-1-0716-3159-1_4,

species [7–9]. More recently, in 2019, the pathogen has been found for the first time in Central Italy (Monte Argentario–Tuscany) on species of Mediterranean maquis and ornamentals [10–12].

In Europe, *Xylella fastidiosa* is regulated as a quarantine organism and its introduction into and movement within the Union territory is prohibited. For these reasons new tools for the rapid molecular diagnosis of this pathogen are urgently needed.

During the last few years, new methods have been developed to address key issues on detection of plant pathogens. Among them, loop-mediated isothermal amplification of DNA (LAMP) [13] is widely used for the diagnosis of pathogens because of its potentially accurate, rapid, cost-effective, and convenient amplification method in both laboratory and field conditions [14–17].

LAMP involves the optical excitation and detection of a pathogen's DNA in an environmental sample mixed with a fluorescent dye as it is heated and amplified in less than an hour. LAMP samples can be processed by relatively low-cost and portable devices, which are available on the market. Due to its rapidity, simplicity, and practicality, LAMP methods have been widely used as an alternative to PCR. The LAMP reaction has high specificity and can be performed at a constant temperature (isothermal amplification). Amplification products can be observed or measured using the turbidity of the reaction mixture [18], or by using portable instrument, allowing a real-time visualization of products on the high-resolution display [17].

Early diagnosis is part of a decision-making process, which in the case of plant diseases may prevent the diffusion of invasive plant pathogens and assist in their eradication. Significant advantages could be gained from moving testing technology closer to the site of sampling and, thereby, reducing the detection time. To this aim, we tested a reliable, fast, and sensitive diagnostic LAMP assay using a portable instrument for the detection of *X. fastidiosa* in-field. The use of rapid, specific, and sensitive point-of-care methods like the LAMP assays enables phytosanitary services to make immediate management decisions, helping in containing environmental and economic losses.

2 Materials

2.1 Sample Collection

1. Plant samples.
2. Plastic bags.
3. Scissors.
4. Scalpel.
5. Ethanol (75–100%).
6. Disposable gloves.

2.2 DNA Extraction from Plant

1. Plant DNA extraction kit (e.g., plant material DNA extraction kit, OptiGene) or a kit for a rapid DNA extraction protocol in-field (*see* **Note 1**).
2. Pipets and pipet tips with filter (10, 200 and 1000 μL).
3. Disposable gloves.

2.3 Real-Time Portable LAMP Amplification

1. Portable LAMP instrument (e.g., Genie II- OptiGene) (*see* **Note 2**).
2. Plasticware specific for LAMP instrument (*see* **Note 3**).
3. LAMP primers to detect *X. fastidiosa* [19] and the COX endogenous plant gene [20, 21] (Table 1) (*see* **Note 4**).
4. Master mix for isothermal amplification (e.g., Isothermal Master Mix [ISO-001], OptiGene Limited, Horsham, UK), suitable for a rapid DNA amplification (*see* **Note 5**).
5. PCR grade water.
6. Pipets and pipet filter tips.
7. Disposable gloves.

3 Methods

3.1 Sample Collection

1. From each plant, leaves are collected and examined during the season of active growth of the plant, in order to maximize the probability to detect *X. fastidiosa* [22] (Fig. 1) (*see* **Note 6**).
2. For each plant species, 5–10 leaves were collected (*see* **Note 7**) and placed in plastic bags (*see* **Note 8**).
3. Cut each leaf transversally into two half portions: the basal portion (containing the petiole) is used for DNA extraction (*see* **Note 9**), after their reduction in smaller fragments (approximately 2 × 2 mm each), by using a sterile scalpel.

3.2 DNA Extraction

Refer to the plant material DNA extraction kit (e.g., OptiGene) for a detailed description of reagents and manufacturer's protocol (*see* **Note 1**):

1. Place the tissue plant fragments (approx. 100 mg wet weight) (*see* **Note 10**) into a bijoux tube containing 1 mL lysis buffer and appropriate steel beads (1 cm diameter).
2. Cap securely and shake each tube manually for 1 min allowing the disruption and the homogenization of plant tissues (*see* **Note 11**) (Fig. 2).
3. Transfer the lysates in 10 μL of prepared plant material into a vial containing the dilution buffer. Cap securely and mix each tube manually for 1 min.

Table 1
LAMP primers used to detect *Xylella fastidiosa* (on RimM gene) and the COX plant control gene

	Target gene	Primer	Sequence (5′–3′)	Length (bp)	References
X. fastidiosa	Ribosome maturation factor (RimM) gene	XF_F3	TAGAGTCTTGGACTGAGCC	19	Aglietti et al. [19]
		XF_B3	ATCGACCCAGTAATACTCGT	20	
		XF_FLP	AGGAGAACGTAATAACCACGG	21	
		XF_BLP	TCCTGGCATCAATGATCGTAAT	22	
		XF_FIP	CACCATTCAACATGGACTCGGTGCGATCTTCCGTTACCAG	40	
		XF_BIP	CTACGAGACTGGCAAGCGTTCGTACCACAGATCGCTTC	38	
Host plant	Cytochrome oxidase (COX) gene	COX_F3	TATGGGAGCCGTTTTTGC	18	Tomlinson et al. [20]
		COX_B3	AACTGCTAAGRGCATTCC	18	
		COX_FLP	ATGTCCGACCAAAGATTTTACC	22	
		COX_BLP	GTATGCCACGTCGCATTCC	19	
		COX_FIP	ATGGATTTGRCCTAAAGTTTCAGGGCAGGATTTCACTA TTGGGT	43	
		COX_BIP	TGCATTTCTTAGGGCTTTCGGATCCRGCGTAAGCATCTG	39	

Fig. 1 Symptoms of decline on Mediterranean maquis caused by *Xylella fastidiosa* in Tuscany (Monte Argentario–Grosseto [Italy], red circle)

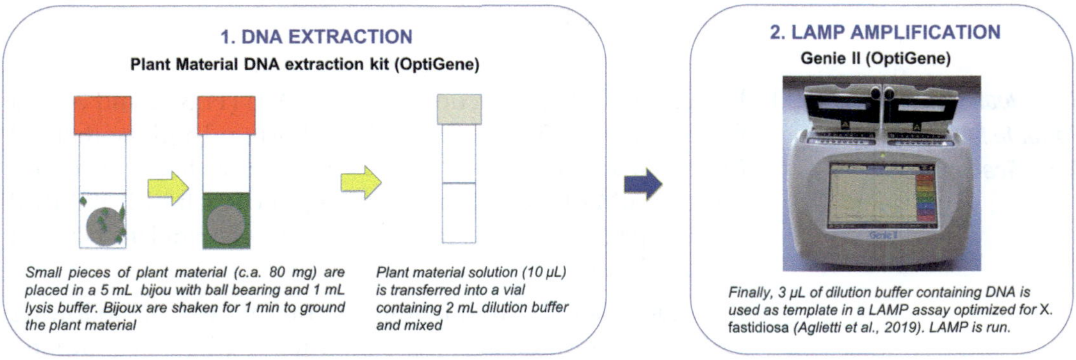

Fig. 2 DNA extraction from plant material and LAMP amplification in-field

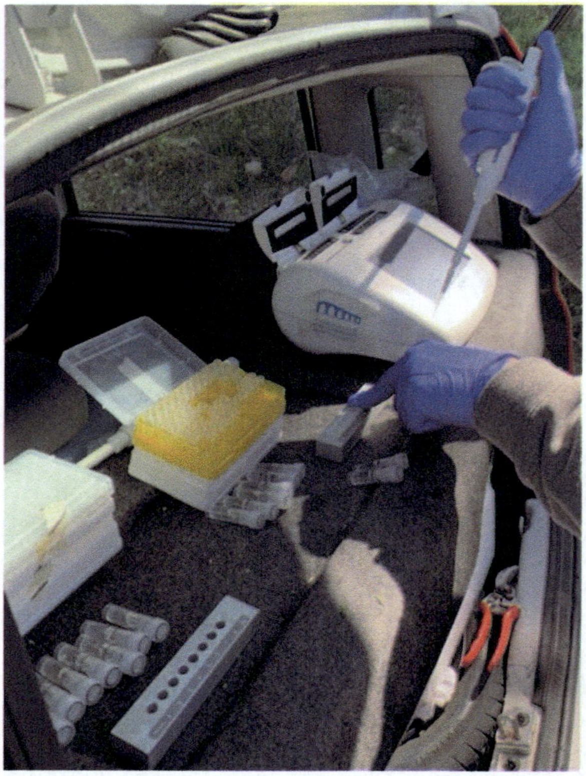

Fig. 3 *Xylella fastidiosa* field diagnosis by using real-time portable LAMP instrument (Genie II, OptiGene)

4. Transfer 3 μL of the prepared plant material in the dilution buffer, and use directly as template in a LAMP assay (*see* Subheading 3.3).

3.3 Real-Time Portable LAMP Amplification

1. Prepare the LAMP reaction (*see* **Note 12**) (Figs. 2 and 3) in a final volume of 25 μL by using the following: 15 μL Isothermal Master Mix; 7 μL LAMP primer mixture (at final concentrations 0.2 μM of each F3 and B3, 0.4 μM of each FLP and BLP and 0.8 μM of each FIP and BIP); 3 μL of template DNA in dilution buffer.
2. Run each sample in duplicate according to the following LAMP protocol: 65 °C for 30 min, followed by an annealing analysis from 98 °C to 80 °C with ramping at 0.05 °C per second that allows the generation of derivative melting curves [23].
3. For each run, include the suitable reference control samples and NTC (no template control, e.g., dH_2O).

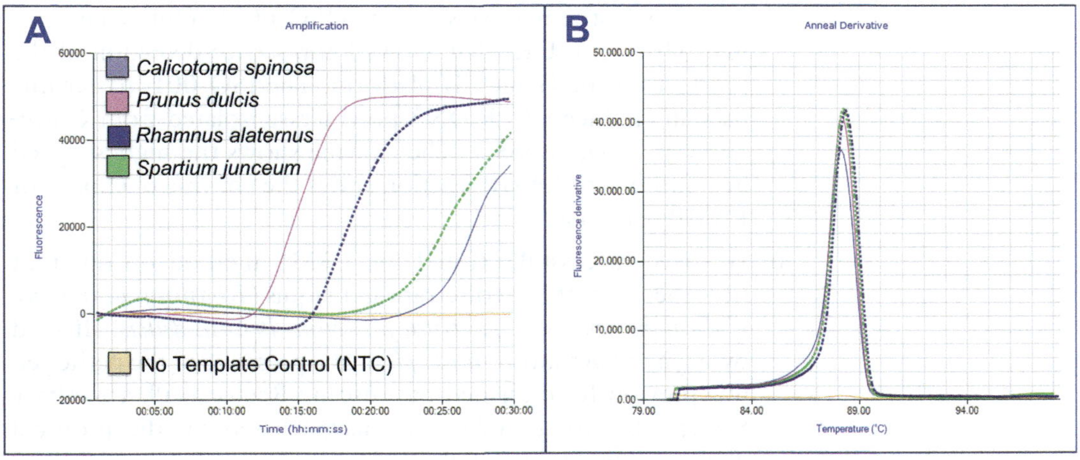

Fig. 4 LAMP detection of *X. fastidiosa*. Selection of kinetic of samples positive to *X. fastidiosa* by using LAMP assay (Genie II) in different host species: (**a**) amplification plot and (**b**) melting curve

4. The LAMP assay is able to recognize *X. fastidiosa* DNA from infected samples, within 30 min of isothermal amplification reaction, with high levels of specificity and sensitivity. An example amplification plot of *X. fastidiosa* and specific melting curve are reported in Fig. 4.

4 Notes

1. We have had success with the OptiGene Plant Material DNA Extraction Kit, which allowed DNA isolation from small amounts of plant tissue (approximately 100 mg). According to the manufacturer's instructions, this kit does not require any sophisticated equipment and can be performed rapidly in the field. Alternatively, other rapid DNA extraction kits can be used in the field.
2. We used the Genie® II portable device (OptiGene), which is suitable for field use and is capable of performing any isothermal amplification protocol that uses target detection by fluorescence measurement.
3. The instrument that we have used has two heating blocks; each block holds a single strip (Genie® strip tubes) that was specifically designed for the device. Each strip has eight tubes with a working volume of 20–150 μL with locking cap providing a closed-tube system.
4. Cytochrome oxidase (COX) gene should be tested for each sample to confirm that host DNA extraction is successfully performed.

5. We used Isothermal Master Mix (ISO-001) (OptiGene Limited, Horsham, UK) that allows fluorescence detection of the product on the Genie® II platform. The ISO-001 master mix uses an engineered LF DNA polymerase isolated from *Geobacillus* sp. SSD, GspSSD. The enzyme shows the highest speed in a fluorescent LAMP reaction and is the enzyme of choice for LAMP reactions.
6. We have successfully tested the LAMP method on different host species of the Mediterranean maquis (*Rhamnus alaternus*, *Calicotome spinosa*, *Spartium junceum*, *Prunus dulcis*) affected by *X. fastidiosa* subsp. *multiplex*. The presence of the pathogen was previously confirmed by Tuscany Regional Phytosanitary Services by using molecular analysis based on the protocol developed by Harper et al. [24].
7. The number of leaves to be collected (include the petiole) is related to the leaf size. For small-size leaves (*C. spinosa*, *S. junceum*, and *R. alaternus*), it is recommended to use a higher number of leaves (e.g., 10), while for large-size leaves (*P. dulcis*), a smaller number (e.g., 5). However, for sampling collection, it is recommended to follow the EPPO guidelines [9] suitable for each host species. The amount of plant tissue to be used is also related to the extraction protocol. In our case we used approximately 100 mg of fresh weight plant tissue.
8. The pathogen may remain on scissors or other tools, possibly infecting the next plant you use them on. For this reason, it is important to sterilize pruning tools with ethanol between different samples to prevent the spread of disease. Label each bag with collected samples with an ID number by using a permanent marker.
9. The portable Genie II instrument is able to analyze 16 samples per run (8 samples in duplicate). For this reason, when eight different bags are collected from the field, the samples can be immediately processed for DNA extraction (*see* Subheading 3.2).
10. *Xylella fastidiosa* colonizes xylem tissue of its hosts, so the petiole and midrib collected from each leaf represent the best source for diagnosis where a higher number of xylem vessels are present [22].
11. Vigorously shake the tube containing the buffer solution and the steel beads, and the plant samples change color to green. This indicates that the plant tissues are properly ground.

12. LAMP mix is previously prepared in the laboratory the same day of the LAMP analysis in the field. The mix is aliquoted in 8-strip tubes (22 μL each tube) and stored in an ice polystyrene box on ice until use in the field. It is recommended to place the strip tubes in a Genie® strip holder (an aluminum holder that can act as a cool block).

Acknowledgments

Authors are grateful to Tuscany Regional Phytosanitary Service for helping to the sampling in the field. This study was funded by European Union's Horizon 2020 Research and Innovation Programme "HOlistic Management of Emerging forest pests and Diseases" (HOMED) (grant no. 771271).

References

1. Almeida RPP, Nunney L (2015) How do plant diseases caused by *Xylella fastidiosa* emerge? Plant Dis 99:1457–1467
2. Bucci EM (2018) *Xylella fastidiosa*, a new plant pathogen that threatens global farming: ecology, molecular biology, search for remedies. Biochem Biophys Res Commun 502:173–182
3. Rapicavoli J, Ingel B, Blanco-Ulate B et al (2018) *Xylella fastidiosa*: an examination of a re-emerging plant pathogen. Mol Plant Pathol 19:786–800
4. Sicard A, Zeilinger AR, Vanhove M et al (2018) *Xylella fastidiosa*: insights into an emerging plant pathogen. Annu Rev Phytopathol 56: 181–202
5. Martelli GP, Boscia D, Porcelli F et al (2016) The olive quick decline syndrome in south East Italy: a threatening phytosanitary emergence. Eur J Plant Pathol 144:235–243
6. Saponari M, Boscia D, Nigro F et al (2013) Identification of DNA sequences related to *Xylella fastidiosa* in oleander, almond and olive trees exhibiting leaf scorch symptoms in Puglia (southern Italy). J Plant Pathol 95:668
7. Denancé N, Legendre B, Briand M et al (2017) Several subspecies and sequence types are associated to the emergence of *Xylella fastidiosa* in natural settings in France. Plant Pathol 66: 1054–1064
8. Olmo D, Nieto A, Adrover F et al (2017) First detection of *Xylella fastidiosa* infecting cherry (*Prunus avium*) and *Polygala myrtifolia* plants, in Mallorca Island, Spain. Plant Dis 101:1820–1820
9. EPPO (2016) EPPO Standards. PM 7/24 (2) *Xylella fastidiosa*. EPPO Bull 46:463–500
10. Saponari M, D'Attoma G, Abou Kubaa R et al (2019) A new variant of *Xylella fastidiosa* subspecies multiplex detected in different host plants in the recently emerged outbreak in the region of Tuscany, Italy. Eur J Plant Pathol 154:1195–1200
11. Marchi G, Rizzo D, Ranaldi F et al (2019) First detection of *Xylella fastidiosa* subsp. *multiplex* DNA in Tuscany (Italy). Phytopathol Mediterr 57:363–364
12. EPPO (2019) *Xylella fastidiosa* subsp. multiplex detected in Toscana region, Italy. EPPO Reporting Service n.01–2019
13. Notomi T, Okayama H, Masubuchi H et al (2000) Loop-mediated isothermal amplification of DNA. Nucleic Acids Res 28:e63
14. Tomlinson JA, Ostoja-Starzewska S, Webb K et al (2013) A loop-mediated isothermal amplification-based method for confirmation of *Guignardia citricarpa* in citrus black spot lesions. Eur J Plant Pathol 136:217–224
15. Hansen ZR, Knaus BJ, Tabima JF et al (2016) Loop-mediated isothermal amplification for detection of the tomato and potato late blight pathogen, *Phytophthora infestans.* J Appl Microbiol 120:1010–1020
16. Panno S, Matić S, Tiberini A et al (2020) Loop mediated isothermal amplification: principles and applications in plant virology. Plan Theory 9:461
17. Luchi N, Ioos R, Santini A (2020) Fast and reliable molecular methods to detect fungal

pathogens in woody plants. Appl Microbiol Biotechnol 104:2453–2468

18. Mori Y, Kitao M, Tomita N et al (2004) Real-time turbidimetry of LAMP reaction for quantifying template DNA. J Biochem Biophys Methods 59:145–157
19. Aglietti C, Luchi N, Pepori AL et al (2019) Real-time loop-mediated isothermal amplification: an early-warning tool for quarantine plant pathogen detection. AMB Express 9:50
20. Tomlinson JA, Dickinson MJ, Boonham N (2010a) Rapid detection of *Phytophthora ramorum* and *P. kernoviae* by two-minute DNA extraction followed by isothermal amplification and amplicon detection by generic lateral flow device. Phytopathology 100:143–149
21. Tomlinson JA, Dickinson MJ, Boonham N (2010b) Detection of *Botrytis cinerea* by loop-mediated isothermal amplification. Lett Appl Microbiol 51:650–657
22. Hopkins DL (1981) Seasonal concentration of Pierce's disease bacterium in grapevine stems, petioles, and leaf veins. Phytopathology 71: 415–418
23. Abdulmawjood A, Grabowski N, Fohler S et al (2014) Development of loop-mediated isothermal amplification (LAMP) assay for rapid and sensitive identification of ostrich meat. PLoS One 9:e100717
24. Harper SJ, Ward LI, Clover GRG (2010) Development of LAMP and real-time PCR methods for the rapid detection of *Xylella fastidiosa* for quarantine and field applications. Phytopathology 100:1282–1288

Chapter 5

Selective Quantification of Chemotropic Responses of *Fusarium graminearum*

Pooja S. Sridhar, Tanya Sharma, and Michele C. Loewen

Abstract

Chemotropism refers to the directional growth of a living organism toward a chemical stimulus. Molecular mechanisms underlying chemotropism of fungal pathogens have recently been enabled by advancements in biological chemotropic assays, with a particular focus on the roles of G-protein-coupled receptors and their plant-derived ligands in chemotropism. Here we describe in detail an assay that enables quantification of chemotropic responses of *Fusarium graminearum*, with variations recently reported for *Fusarium oxysporum* and *Trichoderma atroviride*.

Key words *Fusarium graminearum*, Chemotropism, Plate assay, Microscopy, Chemotropic index, Hyphal length, Directed hyphal growth

1 Introduction

In the first stages of a pathogen-plant host interaction, the pathogen senses its host and either moves (chemotaxis) or grows (chemotropism) toward it [1]. Chemotropism, first documented in the late 1800s [2], specifically refers to the sensing of and directed hyphal growth toward (or away from) a chemical stimulus [3]. Attractant and repulsive chemicals have been found to fall into three categories including nutrient sources, communication signals secreted from other fungi such as pheromones, and compounds secreted from host plants that enable fungal colonization for either symbiosis or pathogenesis. Identification and characterization of such compounds have been enabled over the course of the last century by the application of various qualitative and quantitative biological stimulation assays. Perhaps one of the best representations of the classical assays applied is with respect to the work on *Fusarium graminearum* by Strange and Smith [4] in the early 1970s. In this example, the percent fungal colony (hyphal) extension from a well toward a stimulant was plotted relative to the

Nora A. Foroud and Jonathan A. D. Neilson (eds.), *Plant-Pathogen Interactions*, Methods in Molecular Biology, vol. 2659,
https://doi.org/10.1007/978-1-0716-3159-1_5,

concentration of the stimulant. From this, the stimulant activity of the compound in "units/mL" was obtained, where 1 unit was defined as the amount of stimulant that gives rise to half-maximal extension. Another unique approach for measuring chemotropic growth was described in the 1990s [5], where zoospores of *Phytophthora sojae* were assessed for their ability to respond to isoflavonoid stimuli. In this instance, a chemotropism chamber was created by inverting a coverslip with a 250-microliter suspension of zoospores and placing it over two glass pieces on a glass slide. A capillary tube containing the chemoattractant was then inserted into the chamber and the chemotropic response was measured by counting the number of hyphal germlings that grew into the mouth of the capillary tube.

More recently, molecular and cell biology studies revealed that stimulation of fungal G-protein-coupled receptor family members, by host-derived compounds, mediates most fungal chemotropic responses [6–12]. These findings were largely enabled by improvements to the biological chemotropism assays arising from the advent of digital microscopy. This has enabled quantitative directional and length measurements of hundreds or even thousands of individual hyphae on a daily basis [8] and adaptations of this high-throughput assay to different fungal organisms [9, 13].

Here we expand on the protocol we developed for quantification of the chemotropic responses of *Fusarium graminearum* to various chemical stimulants [9]. We emphasize the relation between chemotropism and the numbers of hyphae growing toward a stimulant (Fig. 1), as well as the relative angle of growth toward the stimulant as an independent indicator of potency (Fig. 2), and the length of hyphae to account for nondirectional growth effects (Fig. 3). We also highlight the changes we made to adapt the original protocol for *Fusarium oxysporum* [8] to *Fusarium graminearum* and compare it to that developed by others [7] with respect to adaptations for application to *Trichoderma atroviride* [13].

2 Materials

All solutions are prepared in UltraPure water (prepared by the reverse osmosis technique to achieve product water resistivity of 18.2 MΩ·cm at 25 °C) using analytical grade reagents. Ensure biological samples are autoclaved before waste disposal. Autoclaving is done at 15 lbs pressure (121 °C) for 15–20 min. Leaving the media in the autoclave for 5–10 min after it has returned to atmospheric pressure is usually a good safety precaution, as superheated liquids can boil over causing both burns and large messes when disturbed.

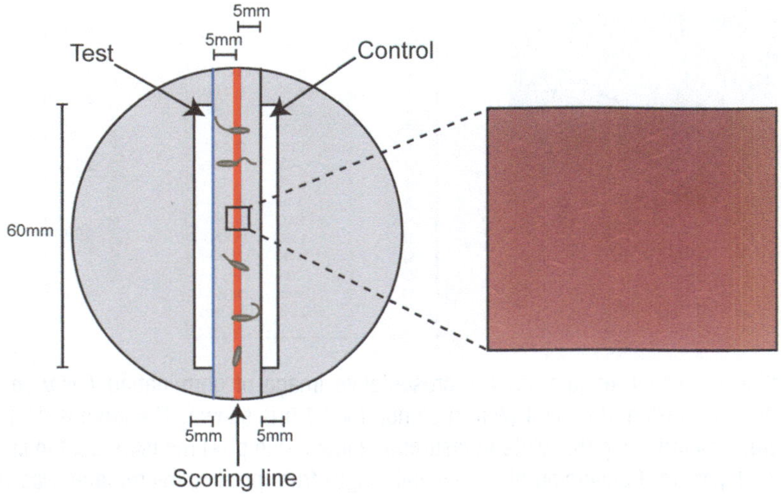

Fig. 1 Schematic representation of quantitative chemotropism plate assay. Chemotropism plate assay used to assess the response of *Fusarium graminearum* toward various stimuli is shown. A central scoring line is shown on Petri plates containing *F. graminearum* conidia in water agar. Two wells are cut in the agar at a 5-mm distance from the scoring line, with each being approximately 60 mm in length. A zoomed-in view of the conidia that fall along the central scoring line is shown, depicting the conidia as well as their germinating hyphae

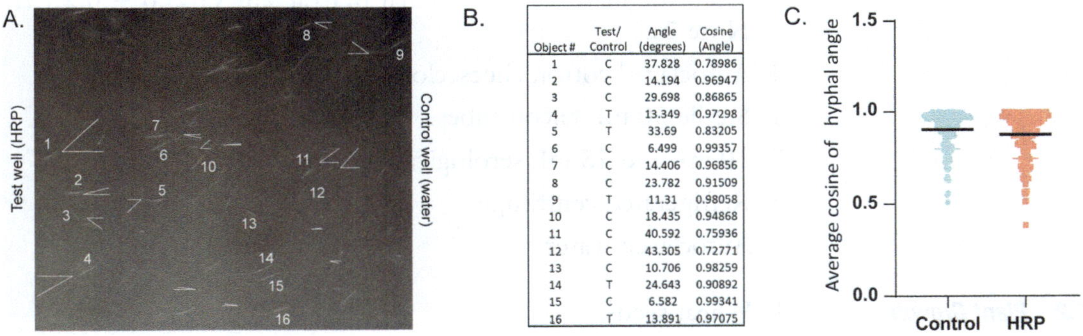

Object #	Test/ Control	Angle (degrees)	Cosine (Angle)
1	C	37.828	0.78986
2	C	14.194	0.96947
3	C	29.698	0.86865
4	T	13.349	0.97298
5	T	33.69	0.83205
6	C	6.499	0.99357
7	C	14.406	0.96856
8	C	23.782	0.91509
9	T	11.31	0.98058
10	C	18.435	0.94868
11	C	40.592	0.75936
12	C	43.305	0.72771
13	C	10.706	0.98259
14	T	24.643	0.90892
15	C	6.582	0.99341
16	T	13.891	0.97075

Fig. 2 Quantification of relative angles of hyphal growth. (**a**) A representative image of germinating *Fusarium graminearum* conidia exposed to test (HRP) and control (water) stimuli for 14 h is shown. Angle measurements are depicted using the white angle symbols. (**b**) Conversion of the angle of hyphal growth into the cosine values. For each germinating conidia labelled in the image, the angle made by its hyphae with respect to the test or control compound is shown in degrees. The cosine of each hyphal angle was determined in Excel. (**c**) The cosine values for all hyphal angles counted ($n = 300$) were plotted as individual data points and the mean is shown by the black bar. Data was plotted in GraphPad 9.0

2.1 Fungal Culturing and Macroconidia Preparation

1. Carboxymethyl-cellulose (CMC) media for fungal conidiation (1 L): 15 g CMC, 1.0 g NH_4NO_3, 1 g KH_2PO_4 monobasic, 0.5 g $MgSO_4.7H_2O$, 1.0 g yeast extract. Dissolve in a 1000-mL glass bottle with stirring. Final pH 7–7.4 (*see* **Note 1**).

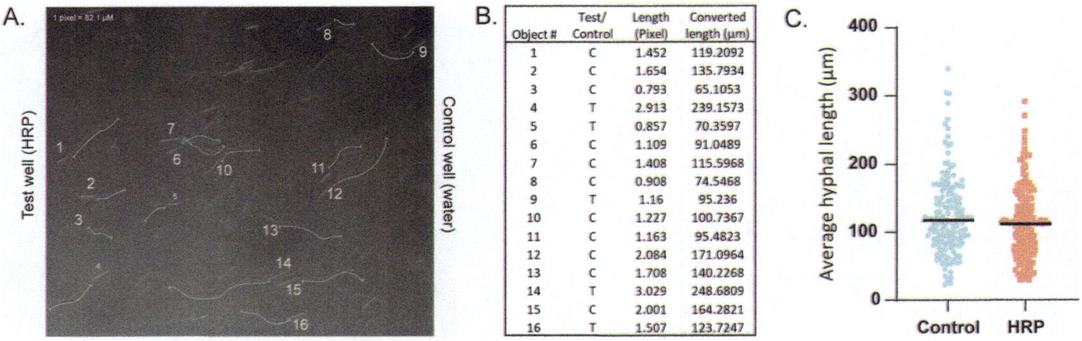

Object #	Test/ Control	Length (Pixel)	Converted length (µm)
1	C	1.452	119.2092
2	C	1.654	135.7934
3	C	0.793	65.1053
4	T	2.913	239.1573
5	T	0.857	70.3597
6	C	1.109	91.0489
7	C	1.408	115.5968
8	C	0.908	74.5468
9	T	1.16	95.236
10	C	1.227	100.7367
11	C	1.163	95.4823
12	C	2.084	171.0964
13	C	1.708	140.2268
14	T	3.029	248.6809
15	C	2.001	164.2821
16	T	1.507	123.7247

Fig. 3 Quantification of hyphal lengths. (**a**) A representative image of germinating *Fusarium graminearum* conidia exposed to test (HRP) and control (water) stimuli for 14 h is shown. The lengths of the hyphae that were measured are depicted using the white measurement lines. One pixel measurement in ImageJ software is equivalent to 82.1 μM. (**b**) Conversion of the hyphal length from pixel to micrometer was determined for each germinating conidia labelled in the image using Microsoft Excel. (**c**) The hyphal lengths for all germinated conidia counted ($n = 300$) were plotted as individual data points and the mean is shown by the black bar. Data was plotted in GraphPad 9.0

2. Potato dextrose agar (PDA) plates for fungal culturing and routine maintenance (1 L): Dissolve 39 g PDA powder in 1.5 mL distilled water. Autoclave and pour media into 100 mm Petri dishes (approximately 20 mL per Petri dish). Store the plates at 4 °C. Final pH (at 25 °C) 5.6 ± 0.2 (*see* **Note 2**).
3. Autoclaved cotton cheesecloth.
4. Sterile 50 mL falcon tubes.
5. 10 mL and 25 mL serological pipettes.
6. Refrigerated centrifuge.
7. Autoclaved water.

2.2 Plant Growth Material

1. Wheat seeds.
2. Topsoil.
3. Sand.
4. Potting soil.
5. Plastic pot (15 cm).
6. 20-20-20 fertilizer.
7. Controlled environment growth chamber.

2.3 Chemotropism Plate Assay

1. Standard 100 mm Petri dishes.
2. 0.5% (*w*/*v*) agar dissolved in distilled water and autoclaved (water agar).
3. Scoopula.

4. Ruler.
5. Colored markers to specify scoring line (red), control well (blue), and test well (black).

2.4 Preparation of Chemical Stimuli

1. Prepare individual solutions of 7 mM final concentration pheromones in 50% (*v/v*) methanol in water. Store at 4 °C for up to 1 month.
2. Prepare a solution of 22.74 μM horseradish peroxidase (HRP) in 0.1 M phosphate-buffered saline (PBS) or distilled water. Store at 4 °C for up to 1 month.
3. Prepare a solution of 200 mM salicylhydroxamic acid (SHAM) in 97% ethanol. Store at 4 °C for up to 1 month.
4. Prepare a solution of 1 mM phenylmethylsulfonyl fluoride (PMSF) in isopropanol. Store at 4 °C for up to 1 month.
5. Prepare a 10 mg/mL stock solution of Proteinase K in water, from lyophilized powder. Store at −20 °C in small aliquots to eliminate repeated freeze-thaw cycles.
6. Centrifugal filters with 3-kDa cutoff.

2.5 Counting Macroconidia and Measuring Hyphal Angles and Lengths Under Microscope

1. Hemocytometer.
2. Coverslips.
3. Stereo light microscope fitted with 20X and 40X objective lenses and a digital camera.

2.6 Installation of Software

1. Image processing and analysis software (*see* **Note 3**).
2. Statistical analysis software, with the ability to plot data and perform unpaired t-tests and one-way ANOVA (*see* **Note 4**).

3 Methods

3.1 Fungal Culture Conditions and Macroconidia Harvest

1. Inoculate 50 mL of liquid CMC in a 125 mL autoclaved Erlenmeyer flask with mycelial plugs or conidial suspensions. For mycelial plugs, culture *F. graminearum* on a PDA plate, and use a sterile spatula to cut out a 1 cm^2 plug of *F. graminearum*. For conidial suspensions, use a pipette to add approximately 20,000 macroconidia from conidial suspensions of *F. graminearum* into 50 mL of liquid CMC in a 125 mL autoclaved Erlenmeyer flask (*see* **Note 5**).
2. Place the flasks in an incubator at 28 °C for 2–5 days with shaking at 170 rpm in the dark.
3. To harvest the conidia, filter the liquid CMC culture through four layers of autoclaved cheesecloth into a 50 mL falcon tube.

4. Centrifuge the filtrate at 3400× *g* for 10 min at 4 °C.
5. Use a sterile serological pipette to remove the media, taking care not to disturb the pelleted conidia.
6. Resuspend the conidia in 50 mL of sterile water, and then centrifuge the suspension at 3400× *g* for 10 min at 4 °C. Discard the supernatant using a sterile pipette.
7. Repeat **step 6**.
8. Resuspended the conidia in 1–3 mL of sterile water and quantify the concentration with a hemocytometer in terms of conidia per mL.
9. Store the harvested conidia at 4 °C for up to 2–3 weeks.

3.2 Wheat Growth Conditions

1. Fill a plastic plant pot with 1:1:1 topsoil, sand, and potting soil and thoroughly water until you see the water flow through the drainage holes.
2. Plant the wheat seeds about 2.5 cm deep into the soil.
3. Place the potted seeds in a plant growth chamber with growth temperature and light conditions of 20 °C during the day, 16 °C at night, with a 16-h photoperiod (750 μmol photons/m^2 × s).
4. Water the plant every second day to saturation and administer "20-20-20" fertilizer solution at a concentration of 2 g/L weekly.

3.3 Quantitative Chemotropism Plate Assay

1. Heat the previously prepared and autoclaved 0.5% (*w/v*) water agar in a microwave until it begins to boil. Ensure all agar has completely melted. Keep aside until the water agar cools down to about 30–35 °C (*see* **Note 6**).
2. Once the agar has cooled, dilute the *F. graminearum* macroconidia to a final concentration of 2.5 × 10^5 conidia per mL in a total volume of 5 mL in a sterile falcon tube (*see* **Notes 7** and **8**). Use a sterile pipette to mix the solution, taking care not to generate air bubbles. Do not vortex the mixture as this will generate air bubbles which cannot be removed easily.
3. Using the pipette, plate the entire 5 mL of mixture onto a Petri dish. Quickly spread the mixture with the pipette evenly over the whole plate, avoiding air bubbles and formation of small chunks of solidified agar.
4. Allow the agar to solidify for about 10 min at room temperature.
5. Once the agar has solidified, turn the plate upside down to draw the scoring, test, and control lines (Fig. 1). Draw the scoring line across the center of the plate using a thick red marker (any color can be used but avoid dark colors). Draw

two parallel lines at a distance of 5 mm on either side of the scoring line using a thin black and blue marker and the test and control compounds, respectively.

6. Use a Scoopula to cut wells in the agar approximately 5 mm wide, 60 mm (or 6 cm) long.
7. Pipette equal volumes (50 μL) of sterile solvent and test compound into the control well and test well, respectively. The nature of the solvent will depend on the stimulus being tested. For example, if the final pheromone solution is in 2.5% (*v*/*v*) methanol, then the control solution should be 2.5% (*v*/*v*) methanol (*see* **Note 9**).
8. Incubate the plates for 14 h at room temperature in the dark (*see* **Notes 10** and **11**).
9. Using a light microscope, count the number of hyphae (Fig. 1) growing toward the test (N_{test}) and control ($N_{control}$) compounds (defined as any hyphae growing at an angle between 0° and 45° toward the respective wells) and calculate a chemotropic index using the following formula:

$$\text{Chemotropic index } (\%) = ((N_{\text{test}} - N_{\text{control}})/(N_{\text{test}} + N_{\text{control}})) \times 100$$

10. Count at least 500 germinated hyphae per plate and repeat the chemotropism plate assay thrice per stimulus.

3.4 Preparation of "Ligand" Containing Samples to Use as Test Compounds

Samples of 50 μL final volume are required for application to the chemotropism assay test well.

1. HRP samples: For HRP alone, dilute an appropriate amount of the stock HRP solution to achieve a working concentration of 4 μM HRP in water at a 50 μL volume and apply immediately to the test well of the chemotropism plate assay. To study the chemoattractive nature of HRP, individually use the following three alternate methods to inactivate HRP and subsequently apply the inactivated solutions to the test well of the chemotropism plate assay. First, mix an appropriate amount of the stock HRP solution with an appropriate amount of the stock SHAM solution, with an appropriate amount of water to produce a 50 μL final volume solution with 4 μM HRP and 60 mM SHAM. Apply directly to the assay. Alternatively, boil a 50 μL 4 μM HRP solution prepared as described above at 95 °C for 10 min in a controlled heat block and then apply to the assay. Finally, mix an appropriate amount of the stock Proteinase K solution to an appropriate amount of the stock HRP solution, with water to produce 50 μL with a final concentration of 1 mg/mL Proteinase K and 4 μM HRP. Incubate for 30 min at room temperature and then apply to the assay. Inhibition of the HRP can be confirmed by peroxidase assay (*see* **Note 12**).

2. α-Pheromone peptides: Dilute stock solutions of both *F. graminearum* (*Fg*) and *S. cerevisiae* (*Sc*) α-pheromone peptides in water to a working concentration of 378 μM with a final volume of 50 μL for each and apply to the assay. To inhibit (proteolyze) the peptides, mix an appropriate amount of the stock Proteinase K solution to an appropriate amount of the stock HRP solution, with water to produce 50 μL with a final concentration of 1 mg/mL Proteinase K and 378 μM pheromone, and incubate the solution at room temperature for 30 min. Stop the reaction adding sufficient PMSF from the stock solution to achieve a final concentration of 1 mM PMSF in the sample.
3. Wheat exudate production: Submerge two flowering wheat heads still attached to the plant in 25 mL of sterile water and incubate them a room temperature for 48 h (*see* **Notes 13** and **14**).

3.5 Hyphal Angle Measurements

1. Use a stereo light microscope to capture images of the germinating hyphae on the chemotropism plates to measure hyphal angles with respect to the gradient of the test compound. Be sure to include a scale on at least one image to use for hyphal length calculations.
2. Open the images in image analysis software (*see* **Note 3**).
3. Hyphal angle measurement to assess consistency of hyphae being included (Fig. 2): Use an angle tool to draw the angle between the hypha and the direction of the gradient of the test and control compounds ($n = 300$). You will obtain values between 0° and 90° with 0° being growth directly toward the test or control wells and 90° being growth toward the side of the plate with no wells. Use a cosine function to determine the cosine of each of the angles measured. Calculate the average of the cosine values of hyphal angles for test and control compounds. This can be used to assess consistency of hyphae being included in the assay.
4. Hyphal angle measurement to determine chemotropic response through angle measurement (Fig. 2): Use an angle tool to trace/obtain the angle formed between the hypha and the gradient of the test compound ($n = 300$). You will measure values between 0° and 180°, with 0° being growth directly toward the test well and 180° being growth toward the control well. Use a cosine function to determine the cosine of each of the angles measured. Calculate the average of all of the cosine values of hyphal angles measured. Repeat this for at least two plates. Perform this analysis for each plate containing different test compounds.

3.6 Hyphal Length Measurements

1. Use a stereo light microscope to capture images of the germinating hyphae on the chemotropism plates to measure hyphal lengths (Fig. 3). Be sure to include a scale on at least one image to use for hyphal length calculations.
2. Open the images in image analysis software (*see* **Note 3**).
3. Use a line tool to draw a line along the length of the hypha and obtain a length measurement in pixels ($n = 300$).
4. Measure the length of the scale in pixels and determine the ratio of pixels to micrometer. This will be used to convert the lengths of hyphae from pixels to micrometer.
5. Calculate the average of all hyphal lengths growing toward test and control compounds.
6. Perform statistical analysis using Student's t-test.

4 Notes

1. CMC is added last as it is harder to dissolve and requires gentle heating (50 °C). Make sure to add it in small quantities and have a screw cap on top to prevent liquid from evaporating. Once dissolved, autoclave the media.
2. Composition of PDA may vary depending on the source of the chemical. In that case, follow manufacturer's instructions for preparing the PDA plates.
3. For measuring the lengths and angles of germinated hyphae, an appropriate image analysis software should be used which has the ability to first process the images, as well as tools to measure parameters such as length and angles. Although there are several freely available as well as commercially available software, we recommend using ImageJ, due to its ease of use and extensive user guides/documentation.
4. Statistical analyses of the chemotropism plate assays and hyphal and length measurements can be done through various software, such as Microsoft Excel, R, SPSS, GraphPad Prism, etc. We used GraphPad Prism because it is user-friendly, but other tools may be used.
5. Do not subculture the conidia more than twice as this could lead to the evolution of mutations in the strain. Revive the strains from glycerol stocks when preparing conidial suspension.
6. Agar needs to have sufficiently cooled before mixing with conidia. The hot agar can damage the conidia, leading to less hyphal germination.

7. The total volume used in the original protocol developed for use in *Fusarium oxysporum* was 4 mL, rather than 5 mL as described in our protocol.
8. We used a concentration of 2.5×10^5 conidia per mL of water agar; however, the concentration of conidia used in the chemotropism plate assay can vary. The original protocol used in *F. oxysporum* [8] as well as for *T. atroviride* [13] describes the usage of 2.5×10^6 conidia per mL of water agar. Another method described the use of 6×10^4 conidia per mL for *F. graminearum* [7].
9. The protocol described for *T. atroviride* used 20 μL of stimulus in each well [13], compared to our volume of 50 μL. If thinner wells are used, less volume of stimulus could be added. However, it is important to make sure that the test or control stimulus completely fills the wells.
10. If plates are incubated too long or at higher temps (i.e., >25 °C), the hyphae become long and intertwined, with multiple hyphae germinating from a single conidium. Some optimization may be required to finalize the best incubation time and temperature for counting of hyphae.
11. Nutrient test compounds tend to also increase hyphal length, making them long and intertwining and more difficult to count—can use a shorter incubation time for these (4–6 h is sufficient to observe the germinating hyphae).
12. Peroxidase activity assay: To assay the activity of peroxidases in the wheat exudate, as well as of HRP, use the previously described assay [14]. Prepare stock solutions of 396 mM pyrogallol (in acetone), 162 mM of H_2O_2 (in water), and 1 μg/mL HRP solution. In a 96-well plate, add the solutions to final concentrations of 13.8 mM pyrogallol, 25 mM H_2O_2, and 20 uL of 1 μg/mL of HRP or wheat exudate. Allow the reaction to incubate at 22 °C for 5 min. Measure the pyrogallol oxidation product (extinction coefficient, ε_{420}, 4400 M^{-1} cm^{-1}) at 420 nm using a plate reader.
13. To measure chemotropism toward wheat, the flowering wheat head still attached to the live plant was placed directly into the test well containing sterile water. Plates were incubated for approximately 14 h at 22 °C in the dark.
14. Other solvents may be used to extract compounds of interest from the wheat plant, such as chloroform [7].

References

1. Turrà D, Nordzieke D, Vitale S et al (2016) Hyphal chemotropism in fungal pathogenicity. Semin Cell Dev Biol 57:69–75
2. de Bary A (1884) Vergleichende morphologie und biologie der pilze, mycetozoen und bacterien. Wilhelm Engelman, Leipzig
3. Miyoshi M (1894) Über chemotropismus der Pilze. Bot Zeitung 52:1–28
4. Strange RN, Smith H (1971) A rapid assay of fungal growth stimulants. Trans Br Mycol Soc 56:485–488
5. Morris PF, Bone E, Tyler BM (1998) Chemotropic and contact responses of *Phytophthora sojae* hyphae to soybean isoflavonoids and artificial substrates. Plant Physiol 117:1171–1178
6. Brown NA, Schrevens S, Van Dijck P et al (2018) Fungal G-protein-coupled receptors: mediators of pathogenesis and targets for disease control. Nat Microbiol 3:402–414
7. Jiang C, Cao S, Wang Z et al (2019) An expanded subfamily of G-protein-coupled receptor genes in *Fusarium graminearum* required for wheat infection. Nat Microbiol 4: 1582–1591
8. Turrà D, El Ghalid M, Rossi F et al (2015) Fungal pathogen uses sex pheromone receptor for chemotropic sensing of host plant signals. Nature 527:521–524
9. Sridhar PS, Trofimova D, Subramaniam R et al (2020) Ste2 receptor-mediated chemotropism of *Fusarium graminearum* contributes to its pathogenicity against wheat. Sci Rep 10:1–15
10. dos Reis TF, Mellado L, Lohmar JM et al (2019) GPCR-mediated glucose sensing system regulates light-dependent fungal development and mycotoxin production. PLoS Genet 15:1–27
11. Johns LE, Goldman GH, Ries LNA et al (2021) Nutrient sensing and acquisition in fungi: mechanisms promoting pathogenesis in plant and human hosts. Fungal Biol Rev 36:1–14
12. Dilks T, Halsey K, De Vos RP (2019) Non-canonical fungal G-protein coupled receptors promote Fusarium head blight on wheat. PLoS Pathog 15:1–26
13. Moreno-Ruiz D, Lichius A, Turrà D et al (2020) Chemotropism assays for plant symbiosis and mycoparasitism related compound screening in *Trichoderma atroviride*. Front Microbiol 11:1–17
14. Maehly AC (1954) The assay of catalases and peroxidases. In: Glick D (ed) Methods of biochemical analysis, vol 1. Interscience Publishers, pp 357–424

Chapter 6

Live-Cell Visualization of Early Stages of Root Colonization by the Vascular Wilt Pathogen *Fusarium oxysporum*

Amey Redkar, Antonio Di Pietro, and David Turrà

Abstract

Fungal phytopathogens induce a variety of pathogenicity symptoms on their hosts. The soilborne vascular wilt pathogen *Fusarium oxysporum* infects roots of more than 150 different crop species. Initial colonization stages are asymptomatic, likely representing a biotrophic phase of infection, followed by a necrotrophic switch after vascular colonization which results in blockage of the plant xylem and killing of the host. Live-cell microscopy techniques have been successfully employed to study interaction events during fungal colonization of root tissues. This technique is widely used to track fungal development during disease progression. Here, we describe a well-established protocol for generation and screening of fluorescently tagged *F. oxysporum* transformants, as well as for live-cell imaging of the early colonization stages of *F. oxysporum* on tomato (*Solanum lycopersicum*) seedlings. The presented experimental design and techniques involved are also applicable to other root infecting fungi.

Key words *Fusarium oxysporum*, mClover3, Live-cell imaging, Tomato, Microscopy, Root

1 Introduction

Fungal pathogens of plants can infect both aerial and belowground organs leading to devastating losses in crop yield [1]. These interactions lead to disease symptoms upon colonization and proliferation of the fungus inside the host plant [2]. Microscopy techniques have been used to study the development of fungal hyphae inside the plant and the nature of defense responses by the colonized host cells [3]. Cytoplasmatic labelling of fungal pathogens with genetically encoded fluorescent tags has several advantages over fixation-based microscopy protocols, including the reduced amount of time

Supplementary Information The online version contains supplementary material available at https://doi.org/10.1007/978-1-0716-3159-1_6.

Nora A. Foroud and Jonathan A. D. Neilson (eds.), *Plant-Pathogen Interactions*, Methods in Molecular Biology, vol. 2659,
https://doi.org/10.1007/978-1-0716-3159-1_6,

required for sample preparation and the possibility to track the infection dynamics of living cells with both spatial and temporal detail [4].

Vascular wilt fungi constitute a particularly destructive group of soilborne pathogens. They attack a broad range of fruit and vegetable crops and are extremely difficult to control, due to their dwelling in the soil or inside the plant vasculature with limited accessibility to chemical treatments [5]. The most notorious causal agent of vascular wilt disease is *F. oxysporum*, a species complex comprising more than 150 *formae speciales* (ff. spp.) attacking different crops [6]. The devastating losses provoked by this pathogen is exemplified by Panama disease of Cavendish bananas, which is caused by the highly aggressive isolate of *F. oxysporum* f. sp. *cubense* tropical race 4 (TR4) [7]. *F. oxysporum* exhibits a hemibiotrophic lifestyle where infection initially involves asymptomatic growth on the living host roots (biotrophic interaction) and eventually switches towards killing of cells and wilting of the entire plant caused by colonization and proliferation of the pathogen in plant xylem (necrotrophic interaction) [6, 8]. Besides being a pathogen on its main host plant species, *F. oxysporum* isolates can also colonize the roots of other plants as endophytes being asymptomatic or even act as biocontrol agents by protecting the plant against pathogenic strains [9–13].

Infection by *F. oxysporum* initiates upon chemotropic sensing of type III peroxidases released by plant roots [14]. After entering the root, Fo grows mainly between the cells of the root cortex [15]. Host tissue penetration and colonization are mediated by a conserved mitogen-activated protein kinase (MAPK) signaling cascade that promotes invasive hyphal growth [16, 17]. During the first stages of plant colonization, *F. oxysporum* secretes a homologue of a plant regulatory peptide known as Rapid ALkalinization Factor (RALF), which promotes infection by triggering alkalinization of the host apoplast [18].

In previous work, *F. oxysporum* fluorescent reporter strains have been used to study the behavior of invasive fungal hyphae and plant responses during early asymptomatic stages of infection and at the onset of plant root colonization [18, 19]. Live-cell imaging with fluorescent strains is particularly advantageous to track fungal infection, considering the low pathogen biomass present during these early stages of colonization. Here, we provide a detailed protocol that will be helpful to study fungal growth and morphogenesis in planta using both wide-field or laser scanning confocal fluorescence microscopy.

2 Materials

All solutions are prepared with sterile distilled UltraPure water (ddH_2O) and stored at room temperature (~25 °C), unless otherwise indicated. Be sure to follow all disposal regulations for waste material.

2.1 Generation of a Fluorescent Protein-Encoding Construct for Fungal Transformation

1. Synthetized codon-optimized fluorophore gene fragments in a plasmid [19] (see below).
2. PCR thermocycler.
3. 200 μL microcentrifuge tubes.
4. Micropipette set.
5. Sterile UltraPure water.
6. PCR components including buffer, dNTPs, $MgCl_2$, primers, and proofreading DNA polymerase (*see* **Note 1**).

2.2 Fungal Cultures

1. The fungal strain used in this example is a derivative of *F. oxysporum* f. sp. *lycopersici* isolate 4287 (FGSC 9935; Fol4287) (Table 1). Fol4287 protoplasts were transformed with a DNA construct encoding three in-tandem copies of the mClover3 fluorophore. Generation of the fluorescent construct is described here, whereas the details of strain generation and transformation are described elsewhere [14, 16]. Strain culture and storage were performed as detailed below.
2. Potato dextrose broth (PDB), prepared according to standard lab protocol.
3. 250 mL Erlenmeyer flasks and magnetic stirrer.
4. Autoclave.
5. Thermoregulated orbital shaking incubator.
6. Autoclavable funnels.
7. Cheesecloth (10 μm mesh).
8. Sterile UltraPure water.
9. 1 mL of 50 mg/mL hygromycin B.
10. 50 mL of 30% (*w*/*v*) glycerol.
11. 50 mL centrifuge tube.

Table 1
***Fusarium oxysporum* strain used in this study**

Strain	Genotype	Gene function	Reference
FGSC 4287-mClover3	*Fol4287:gpdaP*:3xFo-*mClover3*:3xFLAG:SV40PolyA	Cytoplasmatic green fluorescence reporter strain	[19]

2.3 Screening of Fungal Transformants and Live-Cell Imaging

1. 1 mL 95% (*v/v*) perfluorodecalin (PFD).
2. 1 mL of PFD supplemented with 0.005% (*w/v*) calcofluor white (CFW).
3. Gas-permeant poly(dimethylsiloxane) gum.
4. Scalpel.
5. Microscopy slides and coverslips.
6. Sterile UltraPure water.
7. Cell counting chamber.
8. Fluorescence stereomicroscope for low-resolution imaging of putative fluorescent transformants.
9. Microbiological biosafety cabinet.
10. Fluorescence microscope equipped with a scientific camera and a GFP filter set (BP 450/490, FT 510, LP 515).

2.4 Software

1. Microscope-dedicated image acquisition software.
2. ImageJ software [22].

3 Methods

All procedures are carried out at room temperature (~25 °C), unless otherwise specified.

3.1 Generation of Fo-mClover3 Expressing Transformants of F. oxysporum

1. Codon-optimize a genetically encoded fluorescent marker gene based on the *F. oxysporum* f. sp. *lycopersici* codon usage and GC content data retrieved from the Codon Usage Database (http://www.kazusa.or.jp/codon/). In this example the gene for the genetically encoded fluorescent marker mClover3 is used (see plasmid map and sequence in Fig. 1 and Supplemental Fig. 1).
2. Clone the codon-optimized fluorescent marker gene under the control of a strong promoter. In this example, the plasmid gpdaP:3xFo-mClover3:3xFLAG:SV40PolyA containing the pUC57 backbone with the FomClover3 cassette (the *Aspergillus nidulans gpdA* promoter [20], three copies of a *F. oxysporum* codon-optimized mClover3 gene [21] followed by three copies of the FLAG octapeptide tag coding region [3xFLAG], and the *SV40* late polyadenylation signal) is synthesized commercially. This vector is used as a template to amplify the entire FomClover3 cassette using primers gpdA15b (GG ATCCCGAGACCTAATACAGCCCCT) and SV40rev (CTT TTGCTGGCCTTTTGCTCA).
3. Fo-mClover3-labeled fluorescent strains of Fol4287 are obtained by co-transforming fungal protoplasts with the

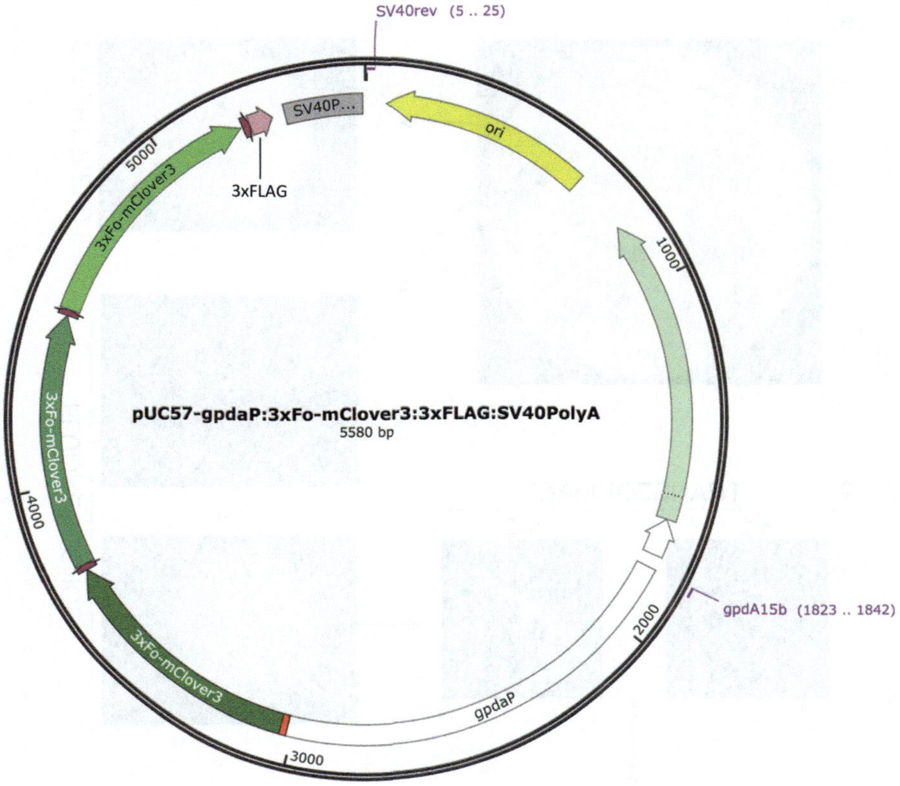

Fig. 1 Structure of the plasmid gpdaP:3xFo-mClover3:3xFLAG:SV40PolyA. Abbreviations used in this map: *gpdaP* promoter region of the *Aspergillus nidulans* glyceraldehyde-3-phosphate dehydrogenase gene, *3xFo-mClover3* three in-tandem copies of a *F. oxysporum* codon-optimized version of the mClover3 fluorophore gene interspaced by 6 aa linker sequences, *3xFLAG,* three in-tandem copies of a *F. oxysporum* codon-optimized version of the FLAG epitope coding sequence, *SV40PolyA* Simian virus 40 early polyadenylation signal sequence

FomClover3 cassette together with the hygromycin B resistance cassette amplified from plasmid pAN7-1 [21] with primers gpdA15b (GGATCCCGAGACCTAATACAGCCCCT) and Trpc8B (AAACAAGTGTACCTGTGCATTC), as previously described [17].

3.2 Screening of the Obtained Fungal Transformants

1. Putative transformants growing on selective media containing 50 μg/mL hygromycin B are imaged under a fluorescent stereomicroscope by cutting a small portion of the transformant colony and imaging it in both bright-field and fluorescence modes to select those showing green fluorescence (Fig. 2). Cytoplasmic Fo-mClover3 expression was observed and quantified for its fluorescence intensity in at least 20 independent transformants using an upright fluorescence microscope equipped with a scientific camera and a GFP filter set (BP 450/490, FT 510, LP 515).

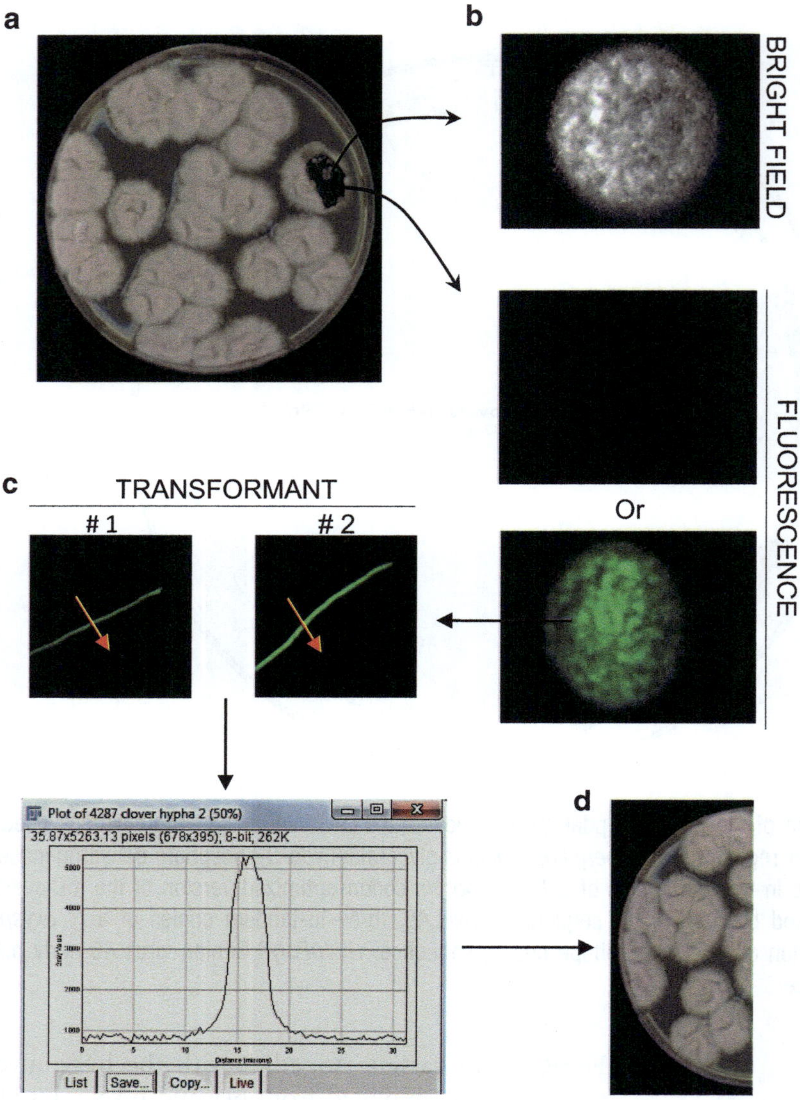

Fig. 2 Schematic view of the selection process of fluorescently tagged *Fusarium oxysporum* transformants. (**a**) Putative transformants grown on selective media containing 50 μg/mL hygromycin B are imaged under a stereomicroscope. (**b**) Micrographs are acquired under a stereomicroscope using bright-field (upper) and fluorescence modes (lower) to preselect the transformants showing green fluorescence. (**c**) Fluorescence microscopy is used to generate high magnification images (630× magnification) of fungal transformant hyphae for ImageJ-based fluorescence quantification and selection of the transformants showing the highest fluorescence values. (**d**) Selected transformants are purified by two successive rounds of single-spore isolation on selective media

2. Fungal transformants showing the highest fluorescence values are identified using the ImageJ software [22]. Briefly, cytoplasmic fluorescence of at least five individual hyphae from each transformant strain is measured by drawing a straight line over the diameter of the hypha followed by the use of the Analyze/

Plot Profile function. Fluorescence values from the different hyphae of each transformant (arbitrary units) are averaged and compared among transformants.

3. Fresh microconidia from the selected transformants were obtained by cutting multiple small fragments from the candidate transformant colonies and inoculating them in 250 mL Erlenmeyer flasks containing 25 mL of autoclaved PDB and supplemented with 20 μg/mL hygromycin B.
4. Incubate fungal cultures for 3–4 days in an orbital shaker at 28 °C and 150 rpm.
5. To collect microconidia, the fungal culture is filtered through four layers of cheesecloth by using a sterile funnel located on top of a 50 mL centrifuge tube. Once the conidia have passed through the cheesecloth, remove the funnel and centrifuge at 8000× *g* for 10 min. Carefully remove the supernatant, resuspend the microconidia in 1 mL sterile ddH_2O, and determine cell density in a cell counting chamber.
6. Purify the transformants by two successive rounds of single-spore isolation on selective media containing 50 μg/mL hygromycin B.
7. Store the final stock of the freshly collected fungal microconidia at −80 °C in 30% (*w*/*v*) glycerol (added in equal volume to the microconidia suspension) or use directly for live-cell imaging.

3.3 Live-Cell Microscopy of Early Root Colonization Stages of F. oxysporum

1. Dip-inoculate the roots of 2-week-old tomato seedlings with 5×10^6 microconidia/mL of the fluorescent Fol4287 strain in sterile ddH_2O. Previously, tomato seeds were sown in vermiculite after surface sterilization in 1% (*v*/*v*) sodium hypochlorite for 30 min and grown in a phyto chamber maintained at 15-/9-h light/dark cycle, 28 °C. Seedlings were uprooted, and roots washed under running tap water, immersed in the spore suspension for 30 min, and repotted in vermiculite after inoculation [23].
2. Collect root samples at the desired time points (usually between 1 and 5 days post inoculation; dpi) (*see* **Note 2**).
3. Prepare 1 mL of a 95% (*v*/*v*) PFD solution and vortex vigorously for 30 s in a microcentrifuge tube (*see* **Note 3**).
4. Cut with a sharp scalpel 2 cm long hand-sectioned portions from the inoculated plant roots (*see* **Note 4**).
5. Transfer one or two root sections to a microcentrifuge tube containing the PFD solution and incubate for 5 min in the dark (wrap the microcentrifuge tubes in aluminum foil).
6. Replace the used PFD solution in the tubes with 1 mL freshly prepared PFD solution (*see* **item 3**) supplemented with 0.005% (*w*/*v*) CFW (*see* **Note 5**).

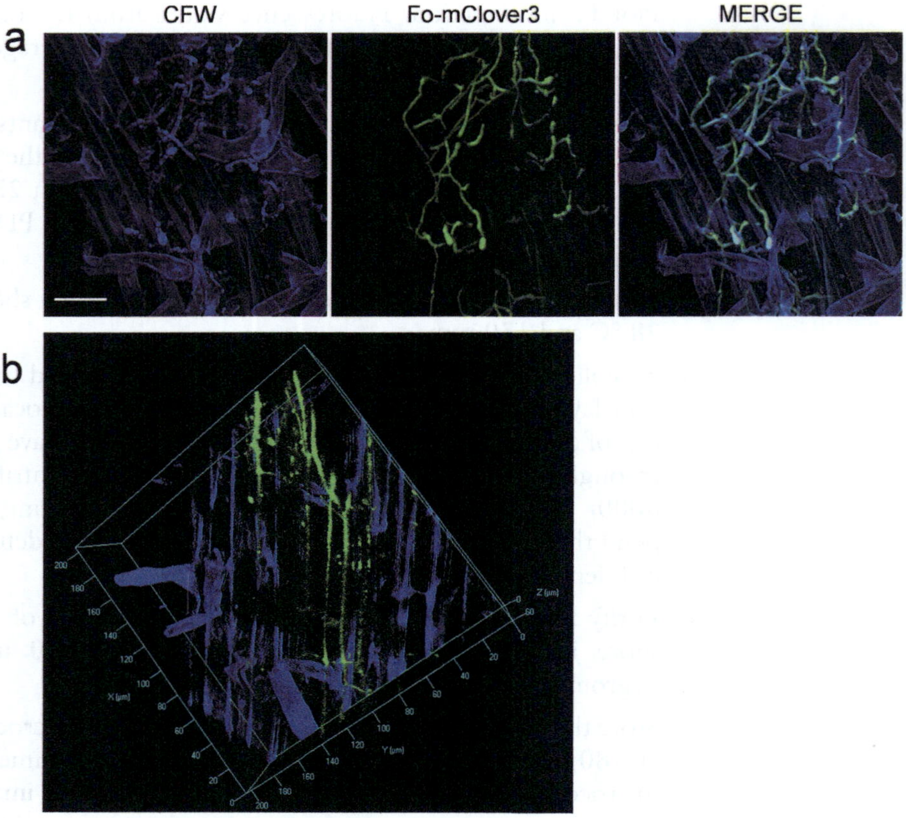

Fig. 3 Representative images of a *Fusarium oxysporum* Fo-mClover3-expressing transformant infecting a tomato root. (**a**) Fluorescence microscopy (400× magnification) showing maximum projections (25 Z-sections) of representative tomato roots stained with calcofluor white (CFW) 6 days after inoculation with a Fo-mClover3-expressing *F. oxysporum* transformant. Scale bar, 50 µm. (**b**) 3D surface rendering showing fungal presence in the inner layers of the same root imaged in (**a**). Note CFW staining of both fungal and plant cell walls in (**a**)

7. Stain root tissues with CFW for 5–10 min in the dark (*see* **Note 6**).
8. Create a squared chamber on a 1 mm-thick microscope slide using a small amount of gas-permeant poly(dimethylsiloxane) gum, as described previously [24].
9. Transfer root sections to the observation chamber filled with an excess of fresh PFD solution (*see* **point 3**). Rapidly seal the chamber by gently and evenly pressing a coverslip onto the poly (dimethylsiloxane) gum.
10. Microscopy is performed using epifluorescence (×400 magnification) and the following filter blocks: CFW staining (G 365, FT 395, LP 420) and GFP (BP 450-490, FT 510, and LP 515) filter set (Fig. 3). Use 20X/0.7NA or 40X/1.2NA objectives (*see* **Note 7**).

11. Data acquisition and image processing (3D reconstruction of Z-series) are performed by using a microscope-dedicated acquisition software and the ImageJ software [22], respectively.

4 Notes

1. It is recommended to use Phusion DNA polymerase from *Pyrococcus* which was used for amplification of all the PCR amplicons generated in this protocol. Alternatively, any proof-reading DNA polymerase can be used.
2. It is recommended to collect the root samples in 1.5- or 2.0-mL microcentrifuge tubes. In case of longer samples, the root can be cut into smaller pieces to fit in a microcentrifuge tube. Usually, the lateral roots are used for microscopy and are sufficient to observe a satisfactory number of fungal colonization events. Moreover, they fit ideally into the microcentrifuge tubes.
3. PFD is a nonfluorescent and nontoxic perfluorocarbon molecule enhancing image resolution in highly autofluorescent samples. Vortex-induced air equilibration of PFD allows for gaseous exchange with the plant tissues.
4. Root sectioning can be performed in a 9 cm petri dish filled with 10 mL of ddH_2O to prevent the tissue from drying out.
5. CFW stains both plant and fungal cell walls through binding of beta-1,3- and beta-1,4 polysaccharides present in plant cellulose and fungal chitin.
6. Alternatively, CFW staining can be replaced with 2 μg/mL of propidium iodide (PI) for 15–30 min in the dark. Longer incubation periods with CFW or PI will allow for better resolution of the inner plant cell layers but may result in higher cytotoxicity in the outer plant cell layers.
7. To image PI-stained roots, the following filter block is used: BP 560/40, FT 585, BP 630/75.

Acknowledgements

This work was supported by the Spanish Ministry of Science and Innovation (MICINN, grant PID2019-108045RB-I00) to A.D.P. A.R. was supported by the European Union's Horizon 2020 research and innovation program under the Marie Skłodowska-Curie grant agreement no. 750669 and by Juan de la Cierva Incorporación grant IJC2018-038468-I from the Spanish Ministry of Science and Innovation MICINN.

References

1. Chaloner TM, Gurr SJ, Bebber DP (2021) Plant pathogen infection risk tracks global crop yields under climate change. Nat Clim Chang 11:710–715
2. Presti LL, Lanver D, Schweizer G et al (2015) Fungal effectors and plant susceptibility. Annu Rev Plant Biol 66:513–545
3. Minker KR, Biedrzycki ML, Kolagunda A et al (2018) Semiautomated confocal imaging of fungal pathogenesis on plants: microscopic analysis of macroscopic specimens. Microsc Res Tech 81:141–152
4. Hemetsberger C, Herrberger C, Zechmann B et al (2012) The *Ustilago maydis* effector Pep1 suppresses plant immunity by inhibition of host peroxidase activity. PLoS Pathog 8:e1002684
5. Berendsen RL, Pieterse CMJ, Bakker PAHM (2012) The rhizosphere microbiome and plant health. Trends Plant Sci 17:478–486
6. Dean R, van Kan JAL, Pretorius ZA et al (2012) The top 10 fungal pathogens in molecular plant pathology. Mol Plant Pathol 13:414–430
7. Ordonez N, Seidl MF, Waalwijk C et al (2015) Worse comes to worst: bananas and Panama disease – when plant and pathogen clones meet. PLoS Pathog 11:e1005197
8. Edel-Hermann V, Lecomte C (2019) Current status of *Fusarium oxysporum formae speciales* and races. Phytopathology 109:512–530
9. Kuldau G, Yates I (2000) Evidence for Fusarium endophytes in cultivated and wild plants. In: Bacon C, White J (eds) Microbial endophytes, vol 5. Marcel Dekker Inc, New York, pp 85–117
10. Fuchs JG, Moënne-Loccoz Y, Défago G (1999) Ability of non-pathogenic *Fusarium oxysporum* Fo47 to protect tomato against *Fusarium* wilt. Biol Control 14:105–110
11. de Lamo FJ, Takken FLW (2020) Biocontrol by *Fusarium oxysporum* using endophyte-mediated resistance. Front Plant Sci 11:37
12. Alabouvette C, Olivain C, Migheli Q, Steinberg C (2009) Microbiological control of soil-borne phytopathogenic fungi with special emphasis on wilt-inducing *Fusarium oxysporum*. New Phytol 184:529–544
13. Veloso J, Díaz J (2012) *Fusarium oxysporum* Fo47 confers protection to pepper plants against *Verticillium dahliae* and *Phytophthora capsici*, and induces the expression of defence genes. Plant Pathol 61:281–288
14. Turra D, El Ghalid M, Rossi F, Di Pietro A (2015) Fungal pathogen uses sex pheromone receptor for chemotropic sensing of host plant signals. Nature 527:521–524
15. Pérez-Nadales E, Di Pietro A (2011) The membrane mucin Msb2 regulates invasive growth and plant infection in *Fusarium oxysporum*. Plant Cell 23:1171–1185
16. Di Pietro A, García-Maceira FI, Méglecz E, Roncero MIG (2001) A MAP kinase of the vascular wilt fungus *Fusarium oxysporum* is essential for root penetration and pathogenesis. Mol Microbiol 39:1140–1152
17. López-Berges MS, Rispail N, Prados-Rosales RC, Di Pietro A (2010) A nitrogen response pathway regulates virulence functions in *Fusarium oxysporum* via the protein kinase TOR and the bZIP protein MeaB. Plant Cell 22:2459–2475
18. Masachis S, Segorbe D, Turrà D et al (2016) A fungal pathogen secretes plant alkalinizing peptides to increase infection. Nat Microbiol 1: 16043
19. Redkar A, Sabale M, Schudoma C et al (2022) Conserved secreted effectors contribute to endophytic growth and multihost plant compatibility in a vascular wilt fungus. Plant Cell 34(9):3214–3232
20. Punt PJ, Oliver RP, Dingemanse MA et al (1987) Transformation of *Aspergillus* based on the hygromycin B resistance marker from *Escherichia coli*. Gene 56:117–124
21. Bajar BT, Wang ES, Lam AJ et al (2016) Improving brightness and photostability of green and red fluorescent proteins for live cell imaging and FRET reporting. Sci Rep 6:20889
22. Schneider CA, Rasband WS, Eliceiri KW (2012) NIH image to ImageJ: 25 years of image analysis. Nat Methods 9:671–675
23. Di Pietro A, Roncero MIG (1998) Cloning, expression, and role in pathogenicity of *pg1* encoding the major extracellular endopolygalacturonase of the vascular wilt pathogen *Fusarium oxysporum*. Mol Plant-Microbe Interact 11:91–98
24. Kirchhelle C, Moore I (2017) A simple chamber for long-term confocal imaging of root and hypocotyl development. JoVE 123:e55331

Chapter 7

Transfection of Barley Leaf Protoplasts with a Fluorescently Tagged Fungal Effector for In Planta Localization Studies in Barley

Ana Priscilla Montenegro Alonso and Guus Bakkeren

Abstract

The isolation and transfection of protoplasts from plant leaves have been routinely used for transient expression and functional studies in model plants. However, current approaches to characterize pathogen effector molecules in a cereal host are inefficient and technically challenging. In this chapter, we describe a protocol to isolate and transfect barley mesophyll protoplasts with a fluorescently tagged fungal effector of the barley smut pathogen *Ustilago hordei*. Tagging of a fungal effector with a fluorescent protein and tracking its localization in cells of its natural host provides insight into its putative in planta localization and helps to narrow down the location of putative host interactors.

Key words Leaf protoplast, *Hordeum vulgare*, Barley, In planta localization, Effector, Microscopy

1 Introduction

Recent years have seen a boom in the application of sequencing tools to identify genes encoding candidate secreted effectors in the genomes of plant pathogens [1]. Most of these effectors, predicted via bioinformatics analyses, still need to be experimentally studied to determine their function and role in pathogenicity. Monocots including cereals are more refractory to many molecular techniques and hence constitute a challenge for the functional characterization of predicted effectors. Traditional methods for in planta expression in cereals include particle bombardment, *Agrobacterium*-mediated, type III secretion system-mediated, and viral-mediated delivery systems [2]. These tools have been used to study the subcellular localization of proteins when linked to fluorescent protein moieties such as GFP or mCherry, but the presence of background fluorescence, caused by autofluorescent plant compounds such as chlorophyll, lignin, or phenolic compounds [3] in plant tissues, can hinder protein visualization during microscopy.

Nora A. Foroud and Jonathan A. D. Neilson (eds.), *Plant-Pathogen Interactions*, Methods in Molecular Biology, vol. 2659,
https://doi.org/10.1007/978-1-0716-3159-1_7,

Protoplasts are plant cells that are released from plant tissue using cell wall degrading enzymes and therefore lack associated cell walls [4]. Compared with the abovementioned transient assays, protoplast transfection allows the study of cell-specific responses at a higher resolution and with a reduced background [4] thus providing an alternative system to study the subcellular localization of plant-pathogen effector molecules. Plant protoplasts have been routinely used for transient assays investigating cellular protein localization [5], protein-protein interactions [6], pathogen-triggered signaling pathways [7], and *R*-gene-mediated cell death [8].

This chapter describes a procedure for the isolation of mesophyll protoplasts from the leaves of barley (*Hordeum vulgare*) seedlings to determine the localization of pathogen effector proteins. This method is based on the protocol for the isolation and transfection of protoplasts from *Nicotiana plumbaginifolia* [9] with some modifications.

2 Materials

All materials to be used should be sterile to avoid microbial contamination.

2.1 Plant Growth

1. Seeds of barley (*see* **Note 1**).
2. Sterile distilled water.
3. 70% ethanol.
4. 1% bleach solution (0.05% final concentration of hypochlorite).
5. Plastic growing trays (21″ × 11″ × 2″) and humidity domes.
6. Soil potting mix.
7. Fertilizer 20-20-20.
8. Environmentally controlled growth chamber.

2.2 Protoplast Isolation

1. Scalpel or razor blade.
2. Petri dishes.
3. 50 mL conical centrifuge tubes.
4. 14 mL round-bottom test tubes with cap.
5. Wide-bore 1 mL pipette tips and pipette.
6. Aluminum foil.
7. 0.2 μm syringe filter.
8. Sterile distilled water.
9. Refrigerator set at 4 °C.

10. 10% bleach plus 0.1% Tween 20 solution.
11. Prepare 100 mL enzyme solution (ES) per tray of ~50 seedlings: 0.6 M mannitol, 10 mM 2-morpholinoethanesulfonic acid (MES) pH 5.7, 1.5% cellulase R-10 (*w/v*), 0.75% macerozyme R-10 (*w/v*), 10 mM $CaCl_2$, 0.1% bovine serum albumin (BSA) (*w/v*). Mix for a minimum of 1 h at room temperature and filter sterilize before immediate use (*see* **Notes 2** and **3**).
12. Orbital platform shaker.
13. Prepare 50 mL 0.5 M sucrose per tray of ~50 seedlings: Dissolve 8.65 g of sucrose to 50 mL with water and filter sterilize.
14. Prepare 100 mL washing (W) solution per tray of ~50 seedlings: 150 mM NaCl, 125 mM $CaCl_2$, 5 mM KCl, 6 mM glucose. Adjust pH to 5.8–6.0 with KOH and filter sterilize.
15. Centrifuge that can accommodate 14 mL tubes and can spin at 250× *g* without brakes.
16. Hemocytometer.
17. Light microscope.

2.3 Protoplast Transfection and Microscopy

1. 0.2 μm syringe filter.
2. Tube rack that allows tubes to be positioned at a 45° angle.
3. Prepare 50 mL MES, mannitol, and $MgCl_2$ (MMG) solution per tray of ~50 seedlings: 4 mM MES pH 5.7, 0.4 M mannitol, 125 mM $MgCl_2$. Filter sterilize before use.
4. Prepare 25 mL mannitol and $Ca(NO_3)_2$ (MaNO) solution per tray of ~50 seedlings: 0.5 M mannitol, 0.1 M $Ca(NO_3)_2$. Adjust pH to 7–9 and filter sterilize.
5. Prepare 5 mL PEG, DMSO, and MaNO (PDM) solution per tray of ~50 seedlings: 40% polyethylene glycol (PEG) 4000 (*w/v*) and 10% dimethyl sulfoxide (DMSO) (*v/v*) prepared in MaNO solution. Filter sterilize, divide in aliquots of 400 μL, and store it at 4 °C for immediate use (*see* **Note 4**).
6. Prepare 50 mL minimal media (MM) solution per tray of ~50 seedlings: 0.7 M mannitol, 10 mM $CaCl_2$, 1 mM KNO_3, 1 mM $MgSO_4$, 0.2 mM KH_2PO_4. Adjust pH to 5.7 and filter sterilize.
7. Confocal laser scanning microscope.
8. Glass bottom dish with a 10 mm microwell and cover glass.

3 Methods

3.1 Preparing Plants for Protoplast Isolation

1. Dehull barley seeds by hand and surface sterilize them as follows: one wash with 70% ethanol for 3 min followed by one wash with 1% bleach for 10 min and several rinses of sterile distilled water (*see* **Note 5**).
2. Add soil potting mix to a tray with appropriate fertilizer and water well. Plant barley seeds (3–4 cm deep) and transfer the trays to a growth chamber at 23 °C, 18-h light, and 6-h dark photoperiod conditions. Keep plants under a humidity dome until seedlings are ready for protoplast isolation (*see* **Note 6**). Generally, 50 seedlings grown in a tray should provide enough protoplasts for transfecting 3–4 constructs.
3. Grow barley plants until they are 7–10 days old. At this stage, the seedlings would have reached ~10–15 cm in height, and in some cultivars, the inner second leaf would be emerging (*see* **Note 7**).

3.2 Isolation of Mesophyll Protoplasts from Barley Leaves

1. Prepare ES, 0.5 M sucrose, W solution, MMG solution, MaNO, PDM solution, and MM solution before starting and store them at 4 °C.
2. Use a sterile scalpel to collect leaves from the stem up to the leaf tip from the barley seedlings from one tray of seedlings (Fig. 1). Transfer the leaves into 50 mL conical centrifuge tubes containing 10% bleach and 0.1% Tween 20 solution and sterilize for 5 min by gentle shaking. Then wash the leaves with sterile water at least three times or until the smell of bleach disappears. Afterwards, keep leaves in water to avoid the tissue from drying out (*see* **Notes 8** and **9**).
3. Place leaves in a petri dish and add enough ES to submerge the leaves. Cut the leaves longitudinally into 0.5–1 mm strips using a sterile scalpel, and immediately place the leaf strips in 14 mL test tubes containing ~10 mL of ES. Incubate the tubes in the dark (by wrapping or covering them with aluminum foil) at room temperature for 4 to 5 h with gentle shaking on an orbital platform shaker (*see* **Notes 10** and **11**).
4. Gently pipette the green solution, containing the isolated protoplasts, into new 14 mL test tubes to a total volume of 10 mL (*see* **Note 12**).
5. Gently washing the leaf strips in the incubated 14 mL tube with a sucrose solution releases more protoplasts. Therefore, add 3 to 5 mL 0.5 M sucrose solution to the original test tube containing the leaf strips, gently swirl, and then transfer this solution to the tubes containing the isolated protoplasts (from **step 4**) to bring the volume up to 10 mL, or transfer it to a new 14 mL test tube (for a final volume of 10 mL).

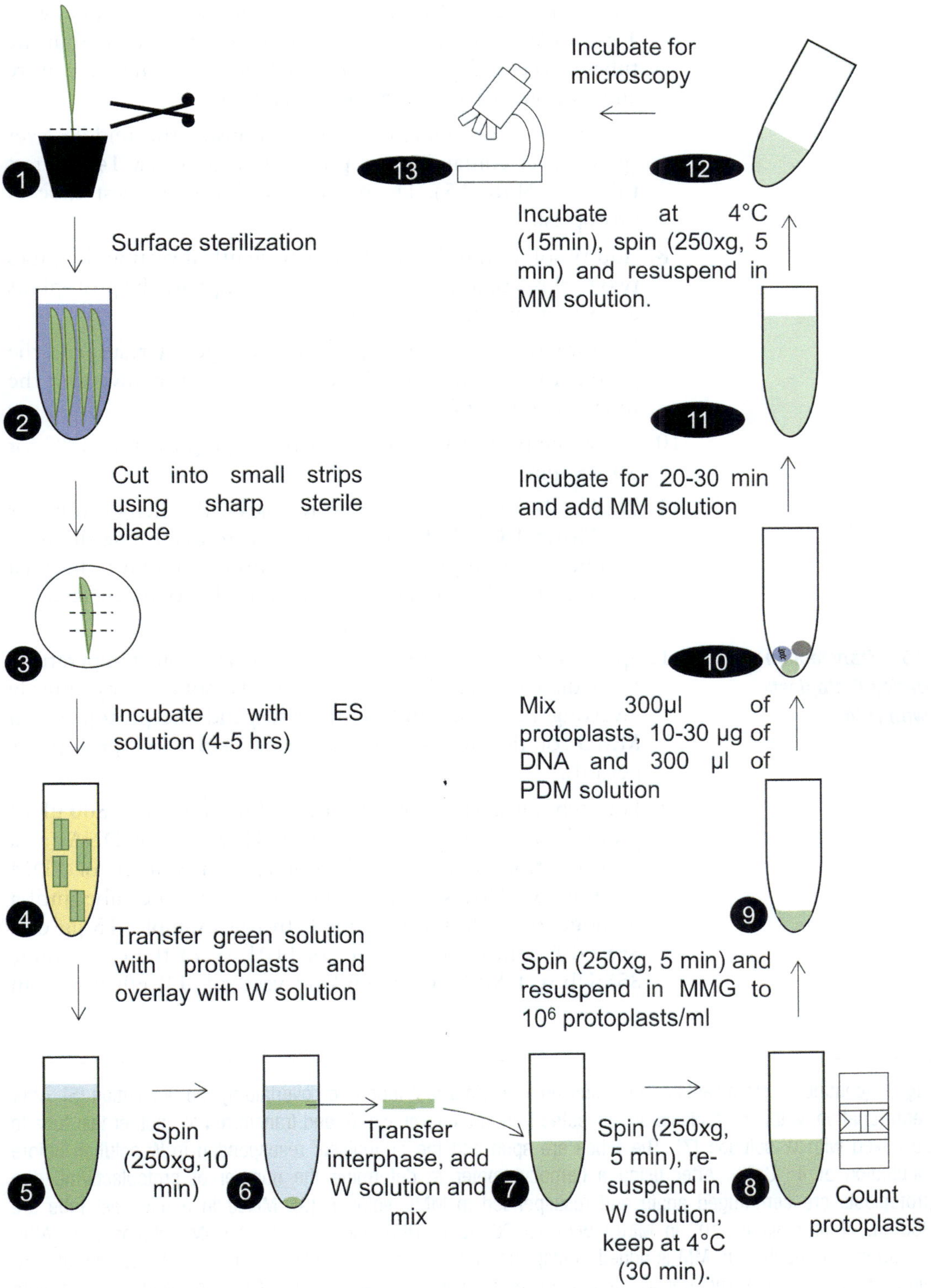

Fig. 1 Graphical representation of the procedure to isolate and transfect protoplasts from barley leaves. Leaves of barley seedlings are harvested [1], surface-sterilized [2], cut into small strips [3], and placed into a 14 mL test tube containing the enzyme solution (ES) [4]. Tubes are incubated for 4–5 h to release the

6. Gently overlay the 10 mL solution in the 14 mL test tubes with 1 mL of W solution. Avoid mixing these two layers. Spin the tubes in a centrifuge at 250× *g* for 10 min, at room temperature and without brakes (*see* **Notes 13** and **14**).
7. Use a wide-bore pipette to gently transfer the middle layer (green layer containing the protoplasts) to a new 14 mL test tube (*see* **Note 15**). Do not exceed 3 mL per test tube of protoplasts.
8. Add W solution to 5 mL and mix by gently inverting the tubes two to three times. Spin the tubes to precipitate the protoplasts at 250× *g* for 5 min (*see* **Note 16**).
9. Remove as much of the supernatant and gently resuspend the protoplasts in 5 mL W solution. Mix by gently inverting the tubes (*see* **Note 17**).
10. Allow the protoplasts to recover by keeping them at 4 °C for 30–45 min.
11. During that time, count the protoplasts with a hemocytometer (*see* **Notes 18** and **19**). It is important to determine the total number of protoplasts available in order to adjust the final volume of MMG solution to be used in the following step.

3.3 Transfection of Barley Protoplasts with DNA

1. Spin protoplasts at 250× *g* for 5 min at room temperature. Immediately after the spin, remove W solution by quickly inverting the tubes and gently resuspend the protoplasts in MMG solution to a concentration of ~4–5 × 10^6 protoplasts per mL.
2. For each transfection, mix in a new 14 mL test tube: 300 μL of protoplasts (~1 × 10^6 protoplasts), 10–30 μg of DNA (at a concentration of 1 μg/μL in water), and 300 μL of PDM solution (*see* **Notes 20** and **21**). Gently mix the tube until a homogenous solution is formed. In this example, 15 μg of a 35S:GFP control (pBIN+ vector [10]) or of the cytoplasmic 35S:UhAvr1-SP:GFP (pK7FWG2 vector [11]) effector from

Fig. 1 (continued) protoplasts and transferred to another test tube to be overlaid by the W solution [5]. After centrifugation, viable protoplasts are collected from the interphase [6] and transferred to another test tube to be mixed with W solution [7]. The tubes are spun and the protoplasts resuspended in W solution before incubation at 4 °C [8]. After using a hemocytometer to determine the number of protoplasts/mL, the protoplasts are centrifuged again and resuspended in MMG solution [9]. Mixed in a new test tube are 300 μL of protoplasts (10^6 in MMG) with 10–30 μg of DNA and 300 μL of PDM solution [10]. After 20–30 min of incubation, MM is added to stop the transfection reaction [11]. Incubation for 15 min at 4 °C allows protoplast recovery after which they are spun down, resuspended in MM solution, and stored at room temperature to allow the protoplasts to express the transfected construct [12]. From 12 h post-transfection, take a sample of protoplasts and place it in a glass-bottom dish to visualize the subcellular localization of a fluorescent protein in confocal microscopy [13]

Ustilago hordei [12] were co-transfected with 15 μg of a transfection control mCherry (pCAMBIA vector [13]) into leaf protoplasts of barley cv. Hannchen (*see* **Note 22**).

3. Incubate each transfection mixture for 20–30 min in the dark at room temperature.
4. Add 10 mL of MM solution to each test tube and gently invert test tubes to mix them well.
5. Allow recovery of the protoplasts by incubating the mixture at 4 °C for 15 min.
6. Spin tubes at 250× *g* for 5 min. Remove supernatant and gently resuspend the protoplasts in ~3–4 mL of MM solution.
7. Incubate the tubes at a 45° angle, in the dark at room temperature until performing further assays (*see* **Notes 23–25**).

3.4 Localization of Fluorescently Tagged Effectors

1. Gently mix the protoplast solution before taking a sample for confocal microscopy assays. Transfer the sample to the microwell of a glass-bottom dish and position a cover glass on top (*see* **Note 26**).
2. Place a water drop under the objective and mount the glass dish into the sample holder of the confocal microscope.
3. To visualize under the confocal microscope a GFP tagged effector and the mCherry transfection control, excite the GFP fluorescence at 480/488 nm and collect the emission at 495–530 nm. mCherry fluorescence is excited at 587 nm and collected at 598–630 nm. Chloroplasts can be visualized using their natural fluorescence by exciting at 480/488 nm and collecting emission at 680–700 nm (*see* **Note 27**). A z-stack of a fluorescing protoplast provides detailed information by collecting images at different focal distances and can be used to computationally compose a 3D image. In this example, the transfected barley protoplasts transfected with the controls (GFP together with a mCherry) showed either or both fluorescence with different intensities (Fig. 2a). The *U. hordei* effector without its signal peptide (SP) and linked at its carboxy terminus to the GFP protein, UhAvr1-SP:GFP [12], showed cytoplasmic localization in barley protoplasts (Fig. 2b).

4 Notes

1. This protocol has been successfully tested using barley cv. Hannchen.
2. Digestion of the cellulose, hemicellulose, and pectin from the mesophyll cells of barley has been successfully achieved by using the cellulase and macerozyme enzymes purchased from Yakult Pharmaceutical, Tokyo, Japan.

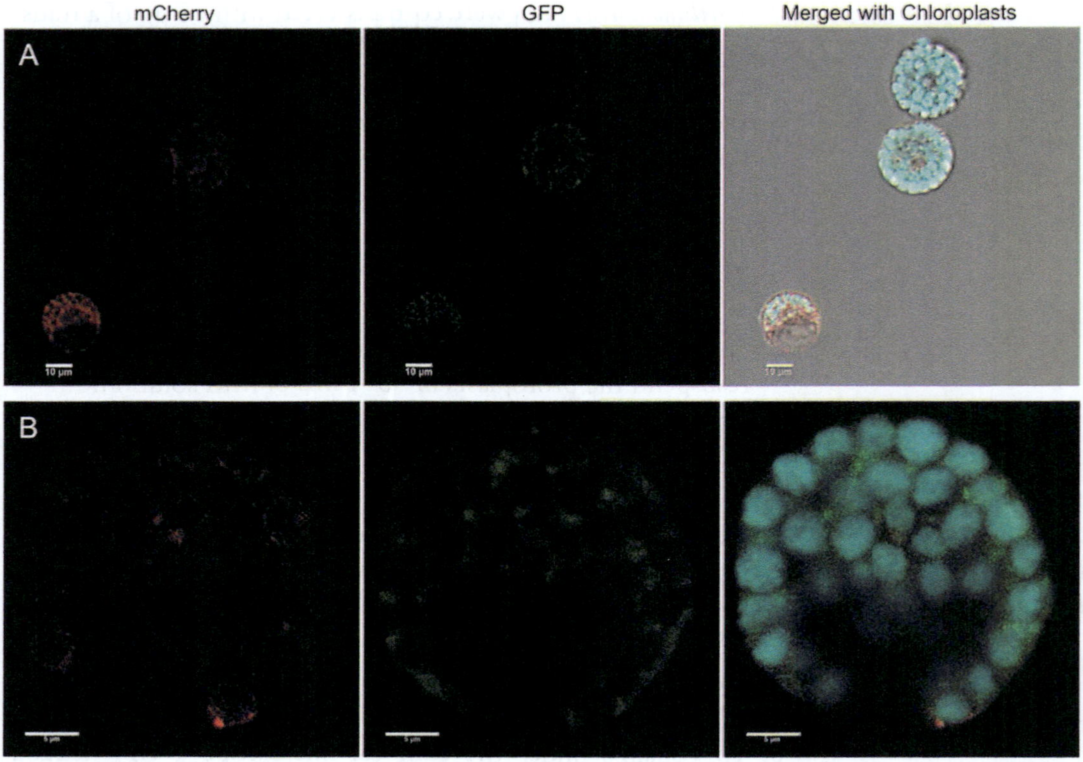

Fig. 2 Barley leaf protoplasts transfected with fluorescently tagged proteins. Protoplasts were isolated from leaves of barley cv. Hannchen and used for co-transfection with a control GFP (**a**) or the UhAvr1-SP:GFP effector (**b**) followed by confocal microscopy. Both constructs (**a** and **b**) were co-transfected together with mCherry to optimize the transfection efficiency. From 14 h post-transfection, GFP (**a**) and UhAvr1-SP:GFP (**b**) showed cytosolic localization. Figure 2**a** and **b** are both single optical sections and the scale bar on them represents 10 μm and 5 μm, respectively

3. Filter sterilization of all solutions described in this chapter was performed using VWR syringe filter units with a pore size of 0.2 microns.
4. The PDM solution is viscous and may need to be pipetted using wide-bore tips. Aliquots can be stored at −80 °C for later use but fresh solution provides better transfection rates.
5. This step is optional but it is recommended to avoid any seed-borne pathogen.
6. Soil should not be dry or overwatered to avoid any stress to the plants.
7. Avoid using older plants to generate protoplasts as they contain a higher amount of wax or secondary metabolites [4]. Also, keeping plants in dark conditions for 1–2 days before commencing the experimental procedure has been shown to reduce the formation of starch granules that can affect the yield and viability of protoplasts [4].

8. Do not harvest barley leaves that are dry, yellow, or damaged for the isolation of protoplasts. Once leaves are harvested, proceed immediately to the following steps to maintain the turgor pressure of the cells and to obtain acceptable levels of protoplast yield and quality.
9. Harvested leaves of barley can also be sprayed with carborundum powder and gently rubbed to generate small abrasions that can facilitate penetration of the enzyme solution and increase the digestion of the barley leaf. Afterwards, rinsing the leaves with water is needed to remove the powder.
10. It is important to use a sharp scalpel or razor blade and avoid tearing of leaf tissue.
11. Protoplasts are very fragile and sensitive to changes in temperature, strong movements (shaking or vortexing), or longer incubation periods, so be gentle throughout the procedure. Gently swirling the test tubes occasionally during the incubation period contributes to the release of protoplasts from the submerged leaf strips.
12. Always use wide-bore pipette tips to transfer protoplasts and to avoid breakage or damage to them. Also, avoid adding solution directly to the protoplasts but instead pipette solutions to the sides of the tube and do not resuspend by pipetting.
13. It is important to use a centrifuge without a brake setting to allow deceleration to occur slowly and thus avoid damage to the protoplasts.
14. Unless indicated, all centrifugation steps are to be carried out at room temperature.
15. After centrifugation, viable and healthy protoplasts (round-shaped with evenly distributed chloroplasts) will be suspended at the interphase between the first and second layer and the bottom of the tube will contain plant debris.
16. **Step 8** is important for washing the protoplasts and removing the remaining enzymes.
17. In **step 9**, it is possible to pool the bottom layer containing protoplasts of several tubes and then complete to 5 mL with W solution.
18. To count in a hemocytometer, place a ~20 μL droplet of the mixed protoplast solution in the hemocytometer counting chamber and place a cover glass. Using a microscope, proceed to count the number of intact protoplasts in the four large outer squares (each square has a 1 mm^2 area with 16 inner grids). If required, dilute samples to facilitate the counting of protoplasts. Calculate the number of protoplasts/mL with the following formula: (no. of protoplast counted in all outer squares/no. of outer squares counted) × 10,000 × (dilution factor).

19. Additionally, the viability of protoplasts can be verified with a microscope by staining the protoplasts with a nonfluorescent dye (e.g., Evan's Blue) or a fluorescent dye (e.g., fluorescein diacetate [FDA]; Thermo Fisher). Viability rates of ~80% are desirable for transfection assays [4].
20. The amount of protoplasts used (300 μL of protoplasts at the concentration of 1×10^6) should be enough for confocal microscopy and to collect samples for other assays such as protein blots.
21. The amount of plasmid DNA to be used for transfection depends on the size and molecular weight of the construct and should be determined experimentally.
22. During initial experiments, it is recommended to include a fluorescent transfection control (e.g., mCherry) and an untransfected control to determine if the experimental conditions and procedures performed during transfection are optimized.
23. Depending on the construct, transient expression can be seen from 12 h onwards and should be empirically determined. However, it is recommended to perform confocal microscopy and collect samples up to 48 h after transfection. Longer incubation time leads to protoplast degradation and loss of fluorescence.
24. Variations between independent experimental repeats are usually found when isolating and transfecting protoplasts. This could be caused by differences in the quality and yield of the isolated protoplasts and plant growth conditions.
25. The protocol given in this chapter has also provided good results for the isolation and transfection of wheat protoplasts.
26. Microscope slides with cover glass should not be used for microscopic observations of protoplasts as they lead to the bursting of protoplasts and the evaporation of the media solution maintaining the protoplasts viable.
27. To visualize more than one different fluorescent protein under a confocal microscope, a sequential scanning mode should be used to avoid or reduce cross-talk between fluorophores with maxima at different wavelengths.

References

1. Dalio RJD, Herlihy J, Oliveira TS et al (2018) Effector biology in focus: a primer for computational prediction and functional characterization. Mol Plant-Microbe Interact 31:22–33
2. Lee W-S, Hammond-Kosack KE, Kanyuka K (2015) *In planta* transient expression systems for monocots. In: Azhakanandam K, Silverstone A, Daniell H, Davey MR (eds) Recent advancements in gene expression and enabling technologies in crop plants. Springer, New York

3. García-Plazaola JI, Fernández-Marín B, Duke SO et al (2015) Autofluorescence: biological functions and technical applications. Plant Sci 236:136–145
4. Lung S-C, Schoor S, Sigurdson D et al (2015) Protoplast isolation and staining. In: Yeung ECT, Stasolla C, Sumner MJ, Huang BQ (eds) Plant microtechniques and protocols. Springer, Cham, pp 197–211
5. Frank S, Hollmann J, Mulisch M et al (2019) Barley cysteine protease PAP14 plays a role in degradation of chloroplast proteins. J Exp Bot 70:6057–6069
6. Ehlert A, Weltmeier F, Wang X et al (2006) Two-hybrid protein–protein interaction analysis in Arabidopsis protoplasts: establishment of a heterodimerization map of group C and group S bZIP transcription factors. Plant J 46:890–900
7. Asai T, Tena G, Plotnikova J et al (2002) MAP kinase signalling cascade in *Arabidopsis* innate immunity. Nature 415:977–983
8. Saur IM, Bauer S, Kracher B et al (2019) Multiple pairs of allelic MLA immune receptor-powdery mildew AVRA effectors argue for a direct recognition mechanism. elife 8:e44471
9. Wieczorek A, Sanfaçon H (1995) An improved method for the generation and transfection of protoplasts from *Cucumis sativus* cotyledons. Plant Cell Rep 14:603–610
10. van Engelen FA, Molthoff JW, Conner AJ et al (1995) pBINPLUS: an improved plant transformation vector based on pBIN19. Transgenic Res 4:288–290
11. Karimi M, Inzé D, Depicker A (2002) GATEWAY™ vectors for *Agrobacterium*-mediated plant transformation. Trends Plant Sci 7:193–195
12. Montenegro Alonso AP, Ali S, Song X et al (2020) UhAVR1, an HR-triggering avirulence effector of *Ustilago hordei*, is secreted via the ER–Golgi pathway, localizes to the cytosol of barley cells during *in planta*-expression, and contributes to virulence early in infection. J Fungi 6:178
13. Beauchemin C, Laliberte JF (2007) The poly (A) binding protein is internalized in virus-induced vesicles or redistributed to the nucleolus during turnip mosaic virus infection. J Virol 81:10905–10913

Chapter 8

A Bioinformatic Guide to Identify Protein Effectors from Phytopathogens

Christopher Blackman and Rajagopal Subramaniam

Abstract

Phytopathogenic fungi are a diverse and widespread group that has a significant detrimental impact on crops with an estimated annual average loss of 15% worldwide. Understanding the interaction between host plants and pathogenic fungi is critical to delineate underlying mechanisms of plant defense to mitigate agricultural losses. Fungal pathogens utilize suites of secreted molecules, called effectors, to modulate plant metabolism and immune response to overcome host defenses and promote colonization. Effectors come in many flavors including proteinaceous products, small RNAs, and metabolites such as mycotoxins. This review will focus on methods for identifying protein effectors from fungi. Excellent reviews have been published to identify secondary metabolites and small RNAs from fungi and therefore will not be part of this review.

Key words Co-expression networks, RNA-seq, Systems biology, *Fusarium graminearum*, Bioinformatics, Protein effectors

1 Introduction

Many effector proteins possess canonical features such as secretory signals that will allow access into the host plant [1]. Secreted proteins are then assessed for "effector-like protein properties" such as size and amino acid composition, including positively charged, cysteine/serine content, and length (<300 amino acids) [2]. Recently, patterns of amino acids such as Ile, Gly, Val, Leu, and Thr, as well as the combination of aromatic and positively charged amino acids, have been used to predict fungal effectors [3]. Cysteine content has also been associated with fungal effectors [4]. The majority of fungal effectors are also under constant selective pressure from the host resulting in higher rates of non-synonymous mutations and a so-called "two-speed" genome [5, 6]. The "two-speed" refers to areas of the genome that are rich in repetitive sequences and transposable elements, serving as a

Nora A. Foroud and Jonathan A. D. Neilson (eds.), *Plant-Pathogen Interactions*, Methods in Molecular Biology, vol. 2659,
https://doi.org/10.1007/978-1-0716-3159-1_8,

hotbed for adaptive evolution compared to the core genome [7, 8]. Rates of non-synonymous mutation can thus be used as criteria to identify novel effectors. These criteria must be evaluated carefully as not all effector proteins possess features that are described. For example, the rust fungus *Melampsora lini* effector AvrM has a length of 314 amino acids, but only possesses a single cysteine residue, and therefore, numbers of cysteine residues in protein do not preclude it being an effector [9]. Some effectors are also likely to be conserved among pathogens, thus residing in the core genome with low recombination rates [10]. To avoid the inadvertent exclusion of potential effectors, one should be cognizant of using strict predefined criteria.

Machine learning approaches for identification of effectors from bacteria have been developed which are now being adapted for use in fungi [11]. Machine learning is a family of statistical learning methods able to identify patterns in complex data sets and separate classes based on features in observed data [2]. Machine learning models are trained on positive and negative datasets that are then applied to the data. As the training sets get more robust, prediction accuracy gets better and reliable. The recent availability of fungal genomes and transcriptomes has increased the training datasets, and thus, identification of effectors is getting much easier [12]. This review provides an outline for a machine learning pipeline to identify fungal putative effector (PE) proteins (Fig. 1).

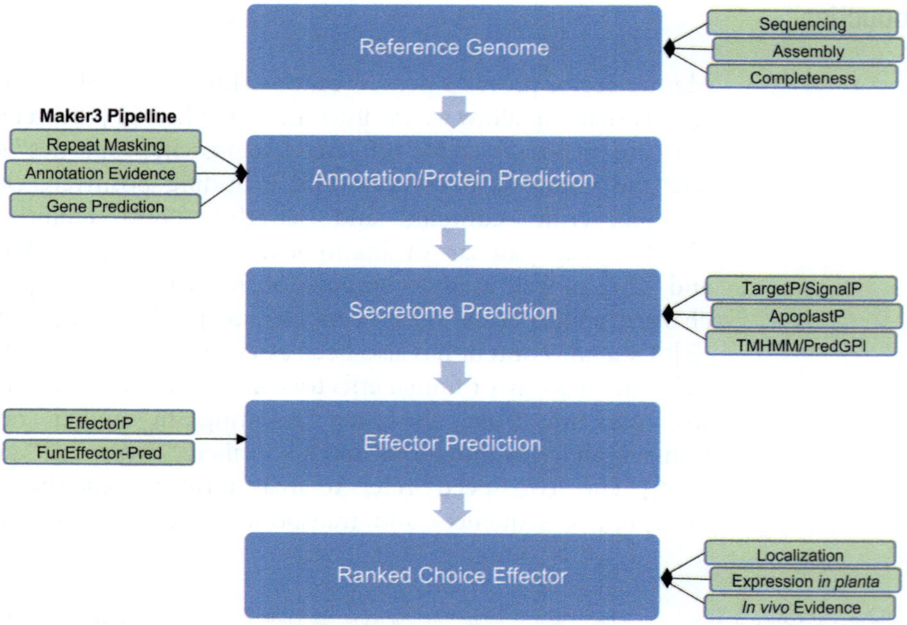

Fig. 1 A workflow to identify effectors from fungi

2 Identification of Putative Effector Proteins

The initial step in identifying PE proteins involves the development of a robust reference genome. With the advent of rapid, high-throughput, and affordable DNA sequencing, genome assemblies for multiple fungi are available (https://www.ncbi.nlm.nih.gov/assembly). Detailed descriptions of genome assembly exist elsewhere and so only a brief description of assembly will be presented here [13, 14]. Genomes can be assembled using a wide variety of open-source pipelines, including SPAdes, ABySS, and Velvet, to name a few [15–17]. Proprietary software is also available such as the CLC Genomics Workbench, which has a built-in assembler (QIAGEN CLC Genomics Workbench 20.0 [https://digitalinsights.qiagen.com/]). A high-quality assembly is crucial to identifying effectors, and toward this end, multiple open-source softwares are available to measure genome completeness, including BUSCO and FGMP [18, 19]. These softwares allow searches for conserved protein and gene elements in the assembled genome and, additionally, reports a completeness score indicating the percentage of conserved features that are absent, fragmented, or complete.

Low-cost sequencing has led to an abundance of publicly available assembled fungal genomes; however, many of these assemblies lack annotation. Annotation is the process of describing genomic features, including gene/coding sequences, untranslated regions, and intron/exon structures, which enable the identification of the proteome. Several open-source annotation softwares are available but care must be taken when annotating a novel genome to avoid common pitfalls, including overestimation of gene content due to repetitive sequences [14]. The MAKER3 software is an open-source annotation pipeline that includes the ability to mask repetitive sequences (RepeatMasker) and provides hints (e.g., expression data, RNAseq) to gene prediction software for more robust predictions [20]. This software utilizes many standalone prediction algorithms including SNAP, Augustus, and GeneMark-ES and allows for evidence-based predictions using external RNAseq experiments, expressed sequence tags (ESTs), and protein homology datasets [21–23]. Annotations predicted by MAKER can be used to determine the full proteome of a given assembly. MAKER3 is available for download on UNIX machines at https://www.yandell-lab.org/software/maker.html.

A secretory signal is a common feature of fungal effectors [1]. Thus, the identification of secreted proteins is a crucial first step with predicted proteome as an input. Proteins are secreted predominantly via the classical ER/Golgi-dependent secretory path, containing well-characterized signal peptides (SPs) for translocating proteins across the membrane [24]. Several tools are used to identify secretory signal peptides, including SignalP and TargetP, accessible at http://www.cbs.dtu.dk/services/SignalP/ and http://www.cbs.dtu.dk/services/TargetP/index.php, respectively (*see* **Notes 1** and **2**) [25, 26]. The latest versions of SignalP (SignalP 5.0) and TargetP (TargetP 2.0) rely on machine learning to predict features in the protein sequences that target organelles, such as signal peptides (Sec), mitochondrial (mTP), chloroplast (cTP), and luminal (luTP) transit peptides [27–30]. An important note is that not all proteins are classically secreted. Nonclassical secretion mechanisms, or leaderless secretion, involve the extracellular translocation of proteins by mechanisms that do not follow the classical ER/Golgi-dependent secretory pathway [31]. Autophagy and extracellular vesicle-dependent pathways play an important role in the nonclassical secretion process and should not be overlooked in effector studies [31]. For example, both the wilting pathogen *Verticillium dahliae* and the soybean pathogen *Phytophthora sojae* contain effector proteins lacking a classical secretory signal [32]. Software such as ApoplastP (http://apoplastp.csiro.au/) can be used to detect unconventionally secreted proteins with great accuracy [33], while software like SecretomeP is also able to predict nonclassically secreted proteins, although it is currently optimized only for mammalian or bacterial species and its use in fungi is questionable [34–37].

Currently, nonclassically secreted proteins targeted to the cytoplasm cannot be computationally identified [10]. As secreted proteins and transmembrane proteins share hydrophobic segments, it is important to screen predicted secreted proteins for transmembrane domains as these may appear as false-positive secreted proteins [12]. One such program, TMHMM, assesses the likelihood for a given protein to contain membrane-spanning alpha-helices [38]. A second program, PredGPI, predicts glycosylphosphatidylinositol (GPI)-anchored proteins—these proteins share similarities to apoplastic proteins leading to false positives [2]. It is important to note that although transmembrane and membrane-bound proteins can result in false positives, transmembrane effectors exist in both pathogenic bacteria and fungi [39, 40]. As an example, in the smut fungus *Ustilago maydis*, a complex of five unrelated effectors and two membrane proteins form a membrane-bound complex that protrudes into the host cell and is critical to virulence [39]. Other extracellular membrane-bound GPI anchor proteins have been implicated in fungal virulence, including GPI7 in *F. graminearum* and *Magnaporthe oryzae* [41, 42]. For this reason, care should be taken when attempting to rule out false-positive, transmembrane proteins [10]. The secretome can be loosely defined as the collection of secreted proteins identified by the TargetP, SignalP, and ApoplastP suite of programs [25, 26, 33].

More recently, EffectorP 3.0 (http://effectorp.csiro.au/), a machine learning fungal effector prediction software, provides a powerful tool to identify fungal effectors from secretome data [43]. A caution to be aware of is that EffectorP relies on positive and negative training datasets. These datasets are often imbalanced—far fewer positive sets are available compared to the negative training sets, leading to misclassification or overrepresentation of effectors [3]. A second fungal effector prediction software, FunEffector-Pred, is designed to learn more effectively from imbalanced datasets [3]. This software also examines novel informative protein features, including combinations of aromatic and positively charged amino acids and patterns of Ile, Gly, Val, Leu, and Thr [3]. FunEffector-Pred is currently only able to run via the terminal from UNIX-based systems. Effectors are critical to the pathogen's ability to modulate host immune response and metabolism and are expected to be expressed during the infection process. In planta expression data can be used to identify PE induced during the early stages of infection. For example, in *F. graminearum,* 88 putative effectors were upregulated in both runner hyphae and infection cushions during colonization of wheat [4]. Additional information on PEs can be gleaned from Wolf PSORT, a bioinformatics tool that predicts subcellular localization of proteins (cytosol, mitochondria, nucleus, chloroplast, etc.) based on their amino acid sequence (https://wolfpsort.hgc.jp/aboutWoLF_PSORT.html.en). Wolf PSORT will augment predictions and pare down PE lists for functional validation purposes.

3 Notes

1. The web-accessible versions of SignalP and TargetP are often heavily trafficked. Consider running the portable UNIX versions on your local machine if you plan on running more than one or two proteomes. Proteomes submitted online will be queued and the results remain available for only 24 h The portable versions may be accessed through https://services.healthtech.dtu.dk/software.php for academic users.
2. TargetP 2.0 uses a deep learning machine learning method to identify N-terminal signals. Deep learning is a form of machine learning using structures modeled after the human brain. This process identifies secretory signals, mitochondrial transit peptide signals, and chloroplast/plastid transit peptide signals and was trained on a dataset of over 10,000 protein sequences and included both positive and negative training sets. Mitochondrial transit peptides (mTP) should be approached carefully as intracellular proteins may be present with mTP signal peptides.

3. For high-throughput analysis of fungal effectors, we recommend the use of local installations via terminal on UNIX-based systems rather than web-based versions. In order to facilitate effector discovery from the command line, Jones et al. [44] have developed an automated pipeline for the prediction and ranking of effector proteins from fungal proteomes [44]. This pipeline can be run in a conda environment allowing for simple installation of dependencies and is freely available (https://github.com/ccdmb/predector).

References

1. Jones DA, Bertazzoni S, Turo CJ et al (2018) Bioinformatic prediction of plant–pathogenicity effector proteins of fungi. Curr Opin Microbiol 46:43–49
2. Sperschneider J, Dodds PN et al (2018) Improved prediction of fungal effector proteins from secretomes with EffectorP 2.0. Mol Plant Pathol 19:2094–2110
3. Wang C, Wang P, Han S et al (2020) FunEffector-Pred: identification of fungi effector by activate learning and genetic algorithm sampling of imbalanced data. IEEE. Access 8 (Ml):57674–57683
4. Lu S, Edwards MC (2016) Genome-wide analysis of small secreted cysteine-rich proteins identifies candidant effector proteins potentially involved in *Fusarium graminearum*-wheat interactions. Gen and Res. https://doi.org/10.1094/PHYTO-09-15-0215-R
5. Alouane T, Rimbert H, Bormann J et al (2021) Comparative genomics of eight *Fusarium graminearum* strains with contrasting aggressiveness reveals an expanded open pangenome and extended effector content signatures. Int J Mol Sci 22:6257
6. Lo Presti L, Lanver D, Schweizer G et al (2015) Fungal effectors and plant susceptibility. Ann Rev Plant Biol 66:513–545
7. Dong S, Raffaele S, Kamoun S (2015) The two-speed genomes of filamentous pathogens: waltz with plants. Curr Opin Genet Dev 35: 57–65
8. Raffaele S, Kamoun S (2012) Genome evolution in filamentous plant pathogens: why bigger can be better. Nat Rev Microbiol 10:417–430
9. Catanzariti AM, Dodds PN, Lawrence GJ et al (2006) Haustorially expressed secreted proteins from flax rust are highly enriched for avirulence elicitors. Plant Cell 18:243–256
10. Depotter JRL, Doehlemann G (2020) Target the core: durable plant resistance against filamentous plant pathogens through effector recognition. Pest Manag Sci 76:426–431
11. Sperschneider J, Gardiner DM, Dodds PN et al (2016) EffectorP: predicting fungal effector proteins from secretomes using machine learning. New Phytol 210:743–761
12. Sonah H, Deshmukh RK, Bélanger RR (2016) Computational prediction of effector proteins in fungi: opportunities and challenges. Front Plant Sci 7:1–14
13. Rice ES, Green RE (2019) New approaches for genome assembly and scaffolding. Annu Rev Anim Biosci 7:17–40
14. Jung H, Ventura T, Sook Chung J et al (2020) Twelve quick steps for genome assembly and annotation in the classroom. PLoS Comp Biol 16:1–25
15. Zerbino DR, Birney E (2008) Velvet: algorithms for de novo short read assembly using de Bruijn graphs. Genome Res 18:821–829
16. Simpson JT, Wong K, Jackman SD et al (2009) ABySS: a parallel assembler for short read sequence data. Genome Res 19(6):1117–1123
17. Bankevich A, Nurk S, Antipov D et al (2012) SPAdes: a new genome assembly algorithm and its applications to single-cell sequencing. J Comp Biol 19:455–477
18. Simão FA, Waterhouse RM, Ioannidis P et al (2015) BUSCO: assessing genome assembly and annotation completeness with single-copy orthologs. Bioinformatics 31:3210–3212
19. Cissé OH, Stajich JE (2019) FGMP: assessing fungal genome completeness. BMC Bioinform 20:1–9
20. Holt C, Yandell M (2011) MAKER2: an annotation pipeline and genome-database management tool for second-generation genome projects. BMC Bioinform 12:491
21. Stanke M, Waack S (2003) Gene prediction with a hidden Markov model and a new intron submodel. Bioinformatics 19(SUPPL 2): 215–225

22. Korf I (2004) Gene finding in novel genomes. BMC Bioinform 5:1–9
23. Lomsadze A, Ter-Hovhannisyan V, Chernoff YO, Borodovsky M (2005) Gene identification in novel eukaryotic genomes by self-training algorithm. NAR 33:6494–6506
24. Peberdy JF (1994) Protein secretion in filamentous fungi – trying to understand a highly productive black box. Trends in Biotech 12: 50–57
25. Almagro Armenteros JJ, Tsirigos KD, Sønderby CK et al (2019) SignalP 5.0 improves signal peptide predictions using deep neural networks. Nat Biotech 37:420–423
26. Armenteros JJA, Salvatore M, Emanuelsson O et al (2019) Detecting sequence signals in targeting peptides using deep learning. Life Sci Alliance 2:1–14
27. von Heijne G (1986) Mitochondrial targeting sequences may form amphiphilic helices. EMBO J 5:1335–1342
28. von Heijne G, Steppuhn J, Hermann RG (1989) Domain structure of mitochondrial and chloroplast targeting peptides. Eur J Biochem 180:535–545
29. von Heijne G (1990) The signal peptide J Membrane Biol 115:195–201
30. Robinson C, Mant A (1997) Targeting of proteins into and across the thylakoid membrane. Trends Plant Sci 2:431–437
31. Miura N, Ueda M (2018) Evaluation of unconventional protein secretion by *Saccharomyces cerevisiae* and other fungi. Cell 7(9):128
32. Liu T, Song T, Zhang X et al (2014) Unconventionally secreted effectors of two filamentous pathogens target plant salicylate biosynthesis. Nat Commun 5:4686
33. Sperschneider J, Dodds PN, Singh KB et al (2017) ApoplastP: prediction of effectors and plant proteins in the apoplast using machine learning. New Phytol 217:1764–1778
34. Bendtsen JD, Jensen LJ, Blom N et al (2004) Feature-based prediction of non-classical and leaderless protein secretion. Protein Eng Des Sel 17:349–356
35. Bendtsen JD, Kiemer L, Fausbøll A, Brunak S (2005) Non-classical protein secretion in bacteria. BMC Microbiol 5:1–13
36. Lonsdale A, Davis MJ, Doblin MS, Bacic A (2016) Better than nothing? Limitations of the prediction tool secretomeP in the search for leaderless secretory proteins (LSPs) in plants. Front Plant Sci 7:1–13
37. Sperschneider J, Williams AH, Hane JK et al (2015) Evaluation of secretion prediction highlights differing approaches needed for oomycete and fungal effectors. Front Plant Sci 6:1–14
38. Sonnhammer ELL, Krogh A (1998) A hidden Markov model for predicting transmembrane helices in protein sequence. Proceedings of the Sixth International Conference on Intelligent Systems for Molecular Biology, 8. papers://4b986d00-906f-493f-a74b-71e29d82b719/Paper/p6291
39. Ludwig N, Reissmann S, Schipper K et al (2021) A cell surface-exposed protein complex with an essential virulence function in *Ustilago maydis*. Nat Microbiol 6:22–730
40. Weigele BA, Orchard RC, Jimenez A et al (2017) A systematic exploration of the interactions between bacterial effector proteins and host cell membranes. Nat Commun 8:532
41. Fernando U, Chatur S, Joshi M et al (2019) Redox signalling from NADPH oxidase targets metabolic enzymes and developmental proteins in *Fusarium graminearum*. Mol Plant Pathol 20:92–106
42. Liu C, Talbot NJ, Chen XL (2021) Protein glycosylation during infection by plant pathogenic fungi. New Phytol 230:1329–1335
43. Sperschneider J, Dodds P (2021) EffectorP 3.0: prediction of apoplastic and cytoplasmic effectors in fungi and oomycetes. Mol Plant-Microbe Interact. https://doi.org/10.1094/MPMI-08-21-0201-R
44. Jones DAB, Rozano L, Debler JW et al (2021) An automated and combinative method for the predictive ranking of candidate effector proteins of fungal plant pathogens. Sci Rep 11: 19731

Chapter 9

Unraveling Plant-Pathogen Interactions in Cereals Using RNA-seq

Bronwyn E. Rowland, Maria Antonia Henriquez, Kirby T. Nilsen, Rajagopal Subramaniam, and Sean Walkowiak

Abstract

Over the past two decades, there have been significant advancements in the realm of transcriptomics, or the study of genes and their expression. Modern RNA sequencing technologies and high-performance computing are creating a "big data" revolution that provides new opportunities to explore the interactions between cereals and pathogens that affect grain yield and food safety. These data are being used to annotate genes and gene variants, as well as identify differentially expressed genes and create global gene co-expression networks. Moreover, these data can unravel the complex interactions between pathogen and host and identify genes and pathways involved in these interactions. This information can then be used for disease mitigation and the development of crops with superior resistance.

Key words Bioinformatics, RNA-seq, Transcriptome, Cereals, Pathogens, Wheat, *Fusarium graminearum*

1 Introduction

Cereals are cultivated grasses that are grown across approximately a billion hectares and contribute to more than half of global agricultural production (www.fao.org/faostat/). Maize, rice, wheat, barley, sorghum, millet, oat, rye, and triticale contribute significantly to the caloric and protein intake of humans and livestock. One of the most challenging production constraints in cereals is the reduction in yield and quality due to more than 30 diseases caused by several diverse genera and species of bacteria and fungi [1]. Many of the microbes that cause these diseases also produce toxins that pose food safety risks. For example, the *Fusarium* head blight (FHB) disease of wheat, which is caused by several species within the genus

---

Bronwyn E. Rowland is the primary author to this chapter.

Nora A. Foroud and Jonathan A. D. Neilson (eds.), *Plant-Pathogen Interactions*, Methods in Molecular Biology, vol. 2659, https://doi.org/10.1007/978-1-0716-3159-1_9,

Fusarium, is estimated to cost the wheat industry >$1 billion in epidemic years in North America alone due to direct yield losses and contamination of grain with fungal toxins [2, 3].

Food security is an important area of ongoing concern given the expected increases in food demand, intensification of crop production, and dynamics in plant pathogen populations. It is estimated that the human population will increase to over 9 billion people by 2050, which will lead to increased food demand [4]. Furthermore, changes in climate are expected to place additional pressures on crop production, resulting in direct physiological and phenotypic changes in crop plants, as well as creating increased abiotic stress from drought and extreme weather [5]. Climate change also directly impacts microbial communities and is predicted to increase the prevalence of diseases caused by crop pathogens that often favor conditions with increased temperature and moisture [6]. There will be a direct impact on plant-pathogen interactions since many diseases occur when the development of both the plant and the pathogen aligns, such as the timing of the discharge of *Fusarium* spores which is coordinated with wheat flowering and leads to FHB [7]. Thus, understanding events during plant-pathogen interactions is critical to meet the current and future food demands, as well as challenges in crop production and disease management.

Most disease management strategies involve an integrated approach of pesticide application, on-farm management such as crop rotations and tilling, and growing disease-resistant cultivars, with the latter being a preferred solution environmentally and economically. However, plant-pathogen interactions are complex and pathogen populations are continuously adapting to overcome crop resistance [8]. As a result, breeders are constantly working to develop new crop varieties that contain diverse sources of resistance. At the same time, disease surveillance programs monitor changes in microbial populations as they adapt to the environment, develop tolerance to pesticides, and overcome host resistance genes [6–8].

Crop breeding and disease surveillance programs have long histories and recent technological advances are revolutionizing both fields. Traditional breeding involves physical inspections of plants in the field, including the assessment of their response to disease pressure. However, next-generation breeding technologies are increasing breeding efficiency by using high-throughput phenomics, genomics, and transcriptomics to both understand and predict disease resistance [9–11]. Similarly, disease surveillance uses these technologies to capture pathogen diversity and predict disease and toxin potential [12, 13]. Much of the focus has been on the application of phenomics and genomics; however, transcriptomics also has significant potential in the study of cereal-pathogen systems.

While the DNA or genome contains the genes that underpin the biology of an organism, the genome is generally considered to be static. Instead, it is the expression of those genes or the ribonucleic acids (RNAs) that controls the organism's morphology and behavior at the cellular level. Each gene has the potential to be transcribed or copied into RNA, which will ultimately be translated to proteins to perform cellular functions. While a gene may only be present as a single copy in the genome, it can be transcribed continuously into several thousand RNA copies in a single cell. In other words, while the genome of the organism is fixed, the RNA is dynamic and responsive to both developmental and environmental cues [11, 14].

Transcriptomics is the study of gene expression and examines global RNA information in a certain cell, group of cells, or tissues. Using transcriptomics, it is possible to understand the behavior of the organism at a molecular level, both temporally and spatially. In a plant-pathogen interaction context, it is often the expression of the genes that determines the pathogenic behavior of the microbe and the defense response of the host that determines disease outcome. As a result, understanding the dynamics of all RNA transcripts, which encompasses the transcriptome, from both the pathogen and the host, is important for understanding why disease occurs and how to prevent it in the future. The low cost and high throughput of RNA sequencing (RNA-seq) technologies have transformed transcriptomics and are an important component of the recent "big data" revolution in biology. Here, we focus on RNA-seq and its application in cereal-pathogen systems and use the interaction between *Fusarium graminearum* and wheat as a model.

2 Recent Advances in RNA Sequencing and Best Practices

2.1 Technologies for Gathering Transcriptome Data, Past and Present

The modern field of transcriptomics emerged from decades of technological developments in molecular biology, such as the polymerase chain reaction (PCR) and the Sanger sequencing methods [15]. Eventually, this led to the conversion of RNA into DNA and the construction of cDNA libraries of expressed sequence tags (ESTs). Quantitative PCR (qPCR), which uses fluorescent dyes to determine absolute and relative amounts of cDNA being amplified, also made it possible to perform targeted quantification of RNA transcripts. In the 1990s, the emergence of microarray technologies allowed for cDNA levels to be detected on a more global scale, increasing the feasibility of large-scale transcriptome studies [16, 17]. Microarrays are chips that contain a library of DNA probes to capture cDNA molecules, which produces a fluorescent signal that is proportional to the amount of cDNA that is present in a sample [18]. However, one of the drawbacks of microarrays is that they can only target sequences that are included as probes on the chip.

In the 2000s, new RNA-seq technologies changed the technological landscape and removed the biases of probe-based transcriptome methods. Early high-throughput short-read sequencing technologies were generating millions of cDNA sequencing reads per run; however, the reads were very short at 35 base pairs (bp). These sequencing reads could then be assembled or mapped to available datasets, such as genome sequences or to EST libraries. Assembly and mapping of RNA-seq reads could then be used to gather counts of the reads mapping to each gene, which would infer gene expression [15]. Over time, short-read sequencing technologies continued to improve both in terms of read length and throughput, giving rise to more modern sequencing platforms such as the Illumina NovaSeq system that is capable of generating billions of paired-end reads with lengths of 2 × 250 bp. When sequencing 250 bp from both ends of a cDNA fragment that is approximately 450 bp long, the small overlap between reads allows them to be merged into a single 450-bp read. These longer reads can be of particular value in the context of cereal transcriptomics because cereal genomes are notoriously complex, having high levels of repetitive elements and gene duplications [9]. Furthermore, cereal crops like wheat are allopolyploid and have multiple highly related subgenomes that hybridized during domestication [10]. Longer sequencing reads reduce the likelihood of reads mapping incorrectly to regions that may have undergone duplication events or are present on multiple subgenomes in the case of allopolyploidy.

The current standard for transcriptomics is RNA-seq and the preferred method to gather RNA count information is short-read sequencing because of its unparalleled throughput. However, it is often desirable/required to gather information on gene structure, alternative splicing, and posttranscriptional modification, which can in some cases be better captured by other sequencing technologies. In the mid- to late 2000s, the development of long-read sequencing technologies by Pacific Biosystems (PacBio) and Oxford Nanopore Technologies (ONT) allowed for the sequencing of cDNA fragments that extend to the entire length of the RNA transcript; this is often referred to as isoform sequencing (Isoseq) [15]. Each of these sequencing technologies offers their own unique benefits. For example, the latest Sequel IIe system by Pacific Biosystems is able to perform iterative sequencing of circularized cDNA molecules, which can generate long reads with accuracies >99% [19]. While ONT reads are less accurate (typically 80–95%), this technology can both sequence cDNA and perform direct sequencing of RNA. The longer reads generated by PacBio and ONT technologies are well suited for the identification of splicing variants or isoforms of genes, as the entire transcript can be captured in a single read. This is not only useful for the annotation of

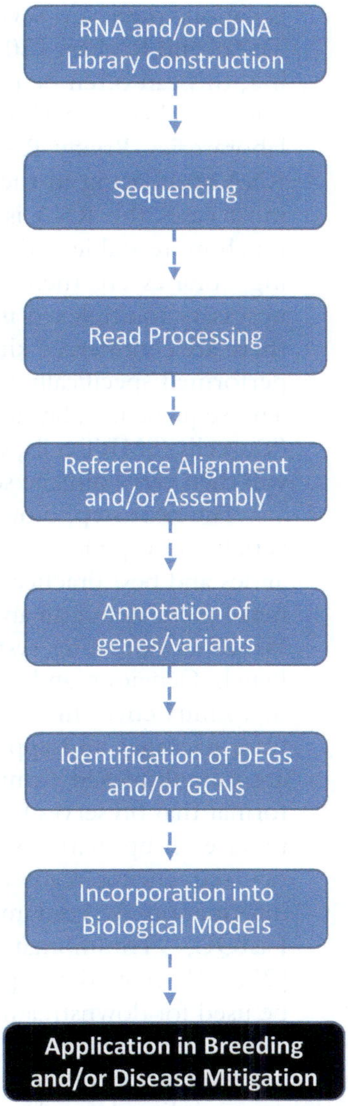

Fig. 1 Typical RNA-seq pipeline for application cereal pathogen systems

genes in genomes, but also capturing how splicing variants change in different cell types and tissues and how they may be changed in response to plant-pathogen interactions.

2.2 RNA-seq Mapping and Assembly

2.2.1 RNA Isolation, Sequencing, and Data Filtering

The typical RNA-seq pipeline involves isolating RNA and preparing the sequencing libraries, sequencing, read processing, read mapping and/or assembly, annotation of genes and/or gene variants, identification of differentially expressed genes (DEGs) and/or global co-expression networks (GCNs), incorporation of the data into biological models, and finally applying the data in breeding or disease mitigation programs (Fig. 1). For RNA isolation, there are several methods and kits available, including TriZol and RNEasy [20]. For the library preparation and sequencing, this can often be performed by a sequencing service provider but can also be performed in most standard molecular biology laboratories. Briefly, RNA can either be used as the input for direct RNA sequencing in the case of certain ONT applications, but in most cases the RNA is reverse transcribed into cDNA, which is much more stable and amenable to diverse applications. Sequencing libraries are then prepared from the RNA or cDNA and if necessary multiplexed if multiple samples are being run together; there are commercial kits available for library preparation, which is performed specifically for the sequencing technology being used. The sequencing libraries are then loaded into instrument-specific flow cells/SMRT cells, which are inserted into the instrument that will then carry out the sequencing reaction.

The next step, bioinformatics and data analysis, is much more variable and application specific; however, there are many commonalities and best practices among pipelines. While software exists to make these steps available to generalists that are most familiar with Windows operating systems (i.e., Galaxy, CLC Genomics Workbench, Geneious, and others), it is favorable to perform the analysis in a Linux environment that can accommodate more diverse software packages and applications [21]. Typically, sequencing reads are demultiplexed from the sequencer and converted into a file format that preserves both the sequence information and the quality at each bp, with fastq being the most common format. These raw sequencing reads are then quality checked and low-quality information or contaminants are removed using tools such as FastQC, Trimmomatic, NanoStat, PoreChip, and others [22, 23]. Once the sequencing reads are processed, they can then be used for downstream analysis specific to the goal of the project.

2.2.2 Read Mapping, Gene Annotation, and Variant Detection

If a reference genome sequence is available, RNA-seq reads can be mapped directly to the reference genome using tools such as Star, HiSat2, or Minimap2 [24–26]. If the genome has not yet been annotated, RNA-seq reads can be complemented with mapping orthologous genes from other related species using tools such as GMAP, or by performing ab initio gene prediction using tools such as GeneMark, Augustus, or FGENESH [27–29]. This will result in multiple overlapping gene calls that can be consolidated using tools such as StringTie2 or Cufflinks [30, 31]. Transcript assembly software is also able to identify alternate transcript isoforms often resulting in greater than one transcript per gene. The gene models can then be validated and the open reading frame predicted using tools such as Transdecoder, GeMoma, and Genomethreader [32, 33]. Predicting gene domains and gene ontology using tools such as InterPro scan can also be useful for inferring gene function and assigning confidence to the gene annotation [34]. Basing annotation on experimental transcript evidence is preferred to using ab initio gene prediction alone. Best practices would dictate that annotation RNA-seq experiments be designed to capture as much of the transcriptome as possible by systematic sampling of multiple growth stages, tissues, and conditions. Similar pipelines have been applied for the annotation of both cereal [10] and pathogen genomes [12, 13].

Using the mapped RNA-seq reads, it is possible to identify a wide array of gene variation. Sequence variants, such as single-nucleotide polymorphisms (SNPs) and insertions/deletions (InDels) between samples, can also be identified based on positional differences in the reads and the reference using tools such as freebayes or the GATK pipeline [35]. Variants can also be sorted and filtered using software such as SnpSift and the impact or effect of the variants on gene function predicted using tools such as SnpEff [36]. In an applied context, these presence/absence and sequence variants and their functional effects can assist in the identification of gene candidates for pathogenicity in the pathogen or disease resistance in the host. For example, an RNA-seq study comparing interactions between *F. graminearum* and susceptible and resistant cultivars of wheat revealed genomic regions in the host that had high rates of gene variation [2]. This variation can be used to develop high-throughput PCR-based assays to determine if these regions are associated with disease resistance in biparental populations or diversity panels and lead to the identification and refinement of quantitative trait loci (QTL). Once QTL are identified, the associated DNA markers can then be used in marker-assisted selection in applied breeding programs to develop superior cultivars with increased disease resistance. For complex traits, such as FHB resistance, which are controlled by many genes and QTL, genome-wide information and genomic selection have also been suggested to predict crop performance and select for superior cultivars [37]. Similarly, RNA-seq of the pathogen can be used to identify variation that may be important for toxin production and/or disease and be used to develop high-throughput PCR-based assays to monitor for this variation in disease surveillance programs [12].

However, using a reference genome for read mapping can be problematic in the detection of variation if either a reference genome is not available or novel sequences are expected, as highly diverged sequences may not map accurately to a reference. In some cases, genome assemblies may also be incomplete or incorrectly assembled in some regions leading to some transcripts being missed in a reference-based approach. RNA-seq reads can alternatively be assembled directly into transcripts using tools such as Trinity [32]. Transcriptome assembly can also be advantageous in the case of cereal genomes that are very large and mostly composed of repetitive intergenic regions. This can reduce the cost of generating high-depth sequencing data of the gene space and allow for the comparison of genes across multiple individuals, or a pan-transcriptome. For example, pan-transcriptome analysis of 368 maize lines identified novel nucleotide-binding leucine-rich repeat (NB-LRR)-containing genes that are receptors potentially involved in disease resistance [38]. One of the challenges of pan-transcriptome studies is that the data is biased towards genes that are actively being expressed. There are other similar DNA-based approaches to enrich for sequencing the gene space, such as exome capture sequencing and RenSeq, which are also able to capture global and gene family-specific data and are not dependent on gene expression, with the latter being specific to capturing and sequencing NB-LRRs [39].

2.3 Analysis of Differentially Expressed Genes

Perhaps the most common use of RNA-seq involves the identification of differentially expressed genes (DEGs) between two experimental conditions. High differential expression is indicative of a gene being either upregulated or downregulated in response to the change in condition. The output of such analyses are lists of genes, their level of differential expression, and statistical significance for being different between conditions. If the goal is to perform analysis of DEGs, it is possible to use the gene models to generate a raw count matrix for each annotated gene using the RNA-seq alignments and tools such as GenomicRanges and IRanges, HTSeq-Count, or featureCounts [40–42]. Factors such as sequencing depth, gene length, and RNA composition can all affect the raw counts in an RNA-Seq dataset. To account for this, raw counts must first be normalized, for which there are several common methods: reads per kilobase million (RPKM), fragments per kilobase million (FPKM), and transcripts per kilobase million (TPM). It is essential that the correct normalization procedure be applied based on the experimental goal. TPM or RPKM/FPKM is typically used to compare the expression of different genes within a single sample. However, when comparing expression levels between multiple different samples, such as in analysis of DEGs, it is often desired to have multiple replicates and assume transcript levels will not follow a typical normal distribution. Statistical packages such as edgeR and DESeq2 are especially suited to normalize count data from replicated samples and perform statistical comparisons between treatments [43, 44]. Both programs use different formulas to perform normalization, but are designed specifically for analysis of DEGs.

Analysis of DEGs between treatments has provided important insights into the behavior of pathogen genes as it pertains to toxin production and disease. As an example, analysis of DEGs between different *F. graminearum* isolates grown in culture and during wheat infection identified genes and pathways involved in secondary metabolism and disease that differs between isolates and treatments, including ~2000 or >14% of genes that are differentially expressed in planta versus in culture [14]. Similarly, an analysis of global expression profiles of *F. graminearum* grown under preferred and non-preferred nutrient conditions was performed, where the latter induces the biosynthesis of the fungal toxin deoxynivalenol (DON), which is a toxin that contaminates wheat grain during infection and poses a food safety risk [45]. The study also revealed that ~2000 of genes were differentially expressed between the two conditions. Closer inspections of genes of interest indicated that many of these genes were involved in the biosynthetic pathway for the production of deoxynivalenol (DON) [45]. A separate study investigating the expression of genes in similar preferred and non-preferred nutrient conditions was performed for mixed cultures where isolates were grown in competition. The results revealed that <100 genes were differentially expressed in response to intraspecies interaction, including genes involved in DON production, which likely translated to reduced accumulation of DON and infection of wheat [12]. Collectively, these and other similar studies provide insights into the molecular mechanisms of toxin production and pathogen infection.

RNA-seq has also been a useful tool for understanding the dynamics of gene expression in the host. For wheat, 850 RNA-seq samples have been combined to form a robust database to describe the dynamics of gene expression, including the expression of homeologous genes of wheat and subgenome-specific responses to different conditions (http://www.wheat-expression.com/) [11]. This database includes multiple conditions and tissues from infection with *Fusarium* that can be compared to a multitude of other conditions, including infection by other pathogens and abiotic stress. Many other studies have also performed more targeted analyses of gene expression during *F. graminearum* infection. When comparing gene expression patterns in wheat during *F. graminearum* infection, there were ~20,000 or 18.5% of wheat genes that were differentially expressed when compared to an uninfected control (Fig. 2) [2]. Many of the differentially expressed genes were located in clusters and had functions that were consistent with interaction with pathogens, disease resistance, and induction of DON production by *F. graminearum* [2]. Expression patterns can also be associated with disease outcomes and be used to identify expression QTL (eQTL) that are genomic regions that contain genes whose expression patterns are consistently associated with disease resistance [46]. This data can be combined with variant information and QTL data to identify gene candidates for disease resistance and be used to select for wheat lines with superior resistance in breeding programs.

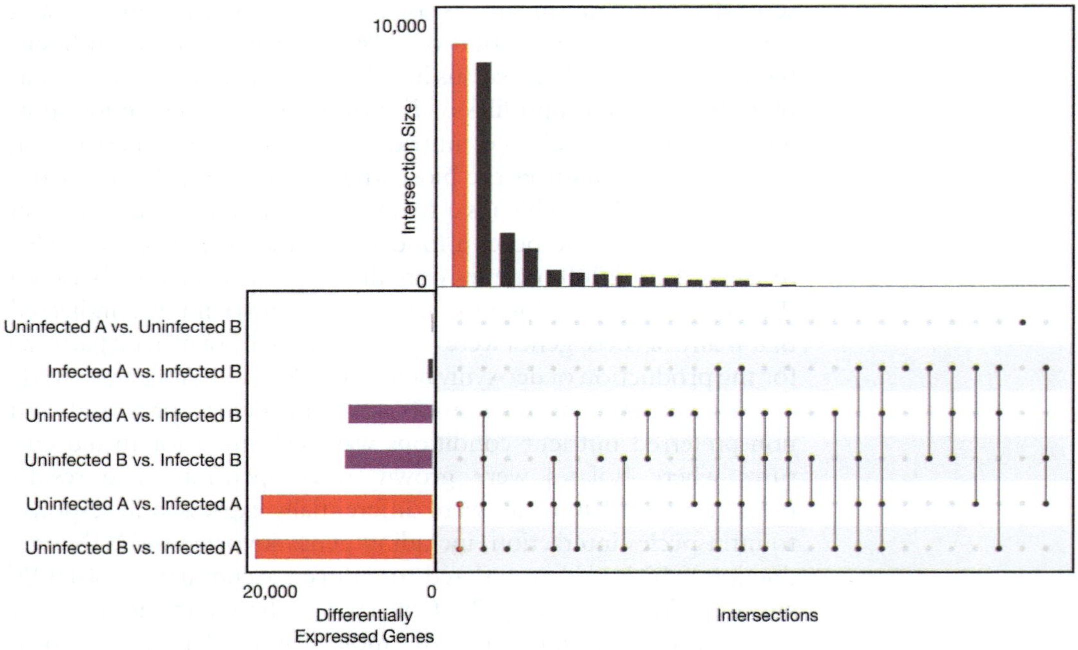

Fig. 2 UpSet plot of intersections between differentially expressed genes in different pairwise comparisons. The susceptible cultivar Roblin was infected with *F. graminearum* or mock inoculated with water. Tissues were then isolated from the inoculated floret (A) and the adjacent floret below the inoculation site (B). Counts for differential expressed genes are indicated by horizontal bars, and counts for intersections/overlap between pairwise comparisons are indicated in vertical bars. The comparisons that are interacting are indicated by dots that are connected by lines. Comparisons and interactions between infected tissues at "A" and the mock inoculated samples are indicated in "red." Purple indicates comparisons between infected tissues at "B" and mock inoculated samples. (Figure adapted from Nilsen et al. [2])

2.4 Advanced Applications of RNA-seq in Co-expression Networks and Systems Biology

Analysis of DEGs is perhaps the most common application of RNA-seq; however, it can be difficult to translate large lists of DEGs and variant information into meaningful biological insights. Many studies inspect or extract individual genes or gene families of interest or genes within genomic regions, such as QTL or eQTL, based on previous knowledge of the pathogen and host. While this can provide additional insights and clarity about known targets, novel data trends are more challenging to identify but should also be explored. One way to explore larger DEG datasets and identify novel data trends is to combine results from multiple comparisons and study their intersections (Fig. 2). Such comparisons can be useful to identify how conditions and comparisons are similar or different and to filter larger lists of genes to smaller lists that may be more biologically consistent or relevant. Gene co-expression network (GCN) analyses are also a new and emerging method that can also bring additional value and context to larger lists of DEGs and are perhaps more practical than looking at intersections of datasets when considering larger numbers of RNA-seq samples simultaneously [47, 48]. To construct a GCN, gene expression data from a biological sample is first filtered and normalized for quality control. Filtering involves removing genes with insignificant expression levels to reduce noise, and normalization involves rescaling the raw data to a fixed range such that expression values are comparable between samples. Next, software such as weighted gene correlation network analysis (WGCNA), ImQCM, or dnapath can be implemented to calculate a co-expression score for each pair of genes in the dataset using a chosen measure of expression similarity [49–51]. For example, partial correlation is a common choice to measure the degree of similarity between two gene expression profiles, and it does so by discounting the influence of the other genes considered confounding variables [51]. A defined threshold for the co-expression score determines which gene pairs have significant co-expression. The GCN therefore connects all significantly co-expressed genes, represented by circular "nodes," using undirected lines or "edges" (Fig. 3a). Network visualization tools such as Cytoscape or GeNeCK are used to display and explore a GCN, as well as to annotate, color, and highlight areas of interest, among other applications [52, 53].

The densely interconnected subnetworks in GCNs known as "modules" (Fig. 3b) can be indicative of genes taking part in a common biological process, either through transcriptional co-regulation, functional relatedness, or by membership to the same protein complex [17]. Co-expression scores also provide additional support to understand the interaction between the gene products within a pathway, with a higher score suggesting a higher likelihood of a direct interaction [51]. In addition to modules, "hubs," or genes with an exceptionally large number of connections are easily identifiable in GCNs (Fig. 3a). Hubs provide

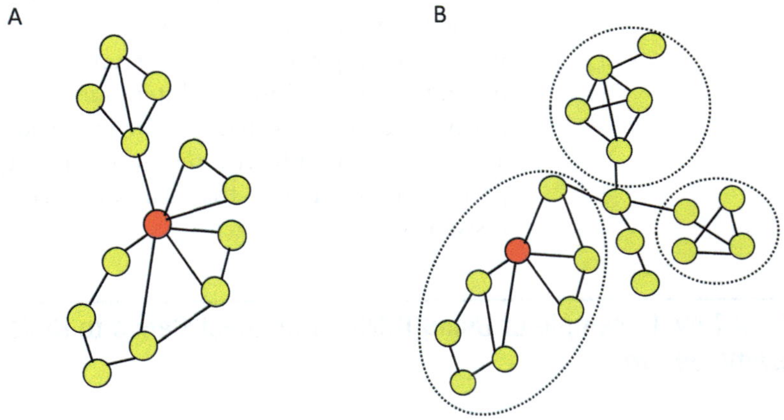

Fig. 3 Gene co-expression networks. (**a**) A small GCN with a single hub gene highlighted in red. (**b**) A small GCN with subnetworks highlighted by dashed ovals. (Figure inspired by Yu et al. [48])

important insight into biological processes as they are more likely to be essential to the pathway [54] and to be conserved throughout evolution and across species [55]. GCNs therefore have the potential to extract more meaningful biological conclusions from genes operating in concert compared to investigating genes on a purely individual level.

Despite the advantages of GCNs, limitations and interpretational challenges exist. Many of these drawbacks stem from the measures used to calculate the similarity of expression between gene pairs as it is unlikely that a single equation, whether it be linear or nonlinear, is able to accurately capture nuanced co-expression relationships. Evaluating and comparing similarity measures is also a challenge as there is no ground truth for which gene pairs should be co-expressed [56]. Additionally, the more common, interpretable measures, such as Pearson's correlation coefficient or partial correlation, are symmetrical measurements and thus do not describe the direction of co-expression. As such, one cannot determine which gene is inducing expression in the other or if another gene altogether is inducing expression in both [57]. Predicting regulatory relationships from co-expression is altogether limited by the very nature of using static transcriptomic samples to understand the inherently dynamic regulation of gene expression [56].

Given the limitations of GCN, it is highly advised to validate hypotheses of gene co-expression and linkage to biological pathways using independent experimental data. Recently, GCN analyses were used to investigate the interaction between wheat and *F. graminearum* [58]. In this study, modules and hub genes were identified in wheat that were involved in FHB resistance and defense, such as response to reactive oxygen species, hormone production, signal transduction, and disease receptors. Many of the genes within the hubs were also co-localized in QTL that were associated with FHB resistance [58]. Using this approach, it is possible to use GCN to generate additional support for candidate genes and regions of interest and their involvement in plant-pathogen interaction. By interconnecting these genes with new pathways and genes that may be involved in related processes, it is also possible to identify new candidate genes involved in plant-pathogen interaction that can be used in breeding for disease resistance.

3 A Novel Example of Using GCNs to Connect Genes to Biological Pathways in *F. graminearum*

To highlight the utility of GCNs, we performed a novel GCN analysis for *F. graminearum* under preferred and non-preferred nutrient conditions using the same RNA-seq data analyzed by Shostak et al. [45]. A differential connectivity network was also

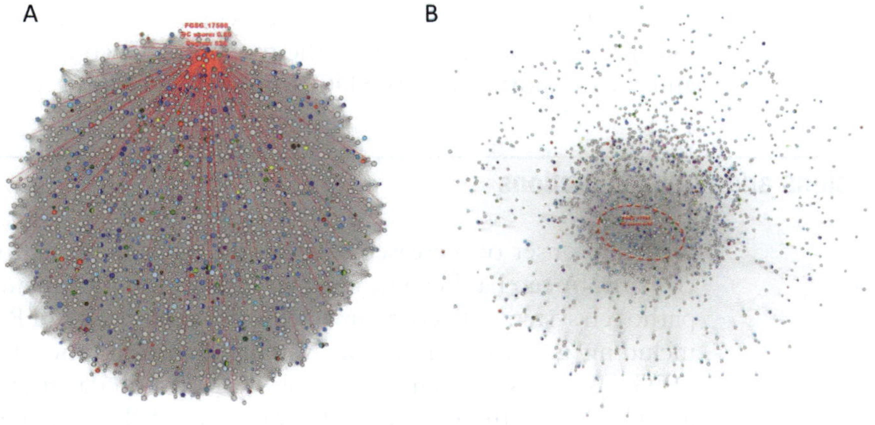

Fig. 4 Differential connectivity network for *F. graminearum* in response to stress. (**a**) FGSG_17598 and its adjacent edges in the global network are highlighted in red. (**b**) FGSG_17598 is a member of a module containing ~250 genes. Nodes are colored by functional description(s): blue is metabolism; green is cellular transport, transport facilitation and transport routes; red is interaction with the environment; purple is cell rescue, defense, and virulence; yellow is energy; brown is protein with binding function or cofactor requirement; and grey is no significant functional description

created to visually compare the GCNs (Fig. 4a), mapping the degree of differential expression of genes under the two conditions as node size and drawing edges between genes with significantly altered co-expression scores. Larger nodes with a relatively larger number of connections therefore indicate genes whose expression is dependent on the condition change and have a wide influence, or hub genes rewired in response to the stress condition. As an example, FGSG_17598 was identified as one of these genes, with a normalized differential connectivity score of 0.80 and a degree of 526 (Fig. 4a). In the study by Shostak et al., this gene was identified as being specifically regulated by Tri6, a key member of the biosynthetic gene cluster that produces DON in *F. graminearum*, under non-preferred nutrient conditions [45]. Additionally, deletion of FGSG_17598 has been shown to increase DON production 14-fold compared to wild-type strains [59]. This putative cytochrome P450 enzyme is therefore likely to be biologically connected to the DON biosynthesis pathway, which is supported by the "detoxification" and "secondary metabolism" functional descriptors linked to this gene based on comparison to gene databases.

One of the modules identified in this GCN encompasses about 250 genes, including FGSG_17598 and Tri6 (Fig. 4b). Five genes with the highest co-expression scores to FGSG_17598 are FGSG_00046 (transporter in the gramillin biosynthetic gene cluster), FGSG_16407 (hypothetical protein), FGSG_04619 (RNA-dependent RNA polymerase-like protein), FGSG_04177 (hypothetical protein), and FGSG_08402 (nitrite reductase).

Future knockout studies of the genes within the identified module may provide more insight into the function of each of these genes and their importance to the DON biosynthesis pathway.

4 Conclusions and Future Directions

With the onset of increased sequencing throughput and reduced costs, RNA-seq has become an emerging tool to understand plant-pathogen systems. There are many diverse applications of RNA-seq, including genome annotation, variant detection, and analysis of DEGs. Bringing biological relevance to RNA-seq data can be challenging due to the volume and complexity of the data. GCNs are especially useful in the age of "big data" as they allow for interpretable exploration of massive datasets composed of thousands of genes and their expression data, simultaneously conveying both local and global trends in gene-gene associations. Linking these trends to prior knowledge, such as genes of interest, QTL, and eQTL, can both provide support for our existing knowledge of the biological processes in the host and pathogen and identify new genes that may be interconnected with these processes. As new technologies and tools continue to develop for transcriptomics, progress in genomics, epigenomics, phenomics, and environomics is also being made, creating a wealth of digital information for cereals, their pathogens, and the physical environments in which they coexist. Linking and modelling these data will not only resolve biological pathways, but also predict the disease potential of pathogens and the performance of crops, leading to the future of disease management.

References

1. Murray TD, Parry DW, Cattlin ND (2009) Diseases of small grain cereal crops: a colour handbook, Softcover edn. Manson Pub, London
2. Nilsen KT, Walkowiak S, Kumar S et al (2021) Histology and RNA sequencing provide insights into Fusarium head blight resistance in AAC Tenacious. Front Plant Sci 11:2114
3. McMullen M, Bergstrom G, De Wolf E et al (2012) A unified effort to fight an enemy of wheat and barley: Fusarium head blight. Plant Dis 96:1712–1728
4. Cole MB, Augustin MA, Robertson MJ, Manners JM (2018) The science of food security. NPJ Sci Food 2:14
5. Raza A, Razzaq A, Mehmood SS et al (2019) Impact of climate change on crops adaptation and strategies to tackle its outcome: a review. Plants 8:34
6. Cavicchioli R, Ripple WJ, Timmis KN et al (2019) Scientists' warning to humanity: microorganisms and climate change. Nat Rev Microbiol 17:569–586
7. Garrett KA, Nita M, De Wolf ED et al (2021) Chapter 24: Plant pathogens as indicators of climate change. In: Letcher TM (ed) Climate change, 3rd edn. Elsevier, pp 499–513
8. Jones JDG, Dangl JL (2006) The plant immune system. Nature 444:323–329
9. Walkowiak S, Gao L, Monat C et al (2020) Multiple wheat genomes reveal global variation in modern breeding. Nature 588:277–283
10. The International Wheat Genome Sequencing Consortium (2018) Shifting the limits in wheat

research and breeding using a fully annotated reference genome. Science 361:eaar7191

11. Ramírez-González RH, Borrill P, Lang D et al (2018) The transcriptional landscape of polyploid wheat. Science 361(6403):eaar6089
12. Walkowiak S, Bonner CT, Wang L et al (2015) Intraspecies interaction of *Fusarium graminearum* contributes to reduced toxin production and virulence. Mol Plant-Microbe Interact 28:1256–1267
13. Walkowiak S, Rowland O, Rodrigue N, Subramaniam R (2016) Whole genome sequencing and comparative genomics of closely related Fusarium Head Blight fungi: *Fusarium graminearum*, *F. meridionale* and *F. asiaticum*. BMC Genomics 17:1014
14. Puri KD, Yan C, Leng Y, Zhong S (2016) RNA-seq revealed differences in transcriptomes between 3ADON and 15ADON populations of *Fusarium graminearum in vitro* and *in planta*. PLoS One 11:e0163803
15. Giani AM, Gallo GR, Gianfranceschi L, Formenti G (2020) Long walk to genomics: history and current approaches to genome sequencing and assembly. Comp Struct Biotechnol J 18:9–19
16. Lenoir T, Giannella E (2006) The emergence and diffusion of DNA microarray technology. J Biomed Discov Collab 1:11–11
17. Weirauch MT (2011) Gene coexpression networks for the analysis of DNA microarray data. Applied Statistics for Network Biology, pp 215–250
18. Schena M (1996) Genome analysis with gene expression microarrays. BioEssays 18:427–431
19. Hon T, Mars K, Young G et al (2020) Highly accurate long-read HiFi sequencing data for five complex genomes. Sci Data 7:399
20. Johnson MTJ, Carpenter EJ, Tian Z et al (2012) Evaluating methods for isolating total RNA and predicting the success of sequencing phylogenetically diverse plant transcriptomes. PLoS One 7:e50226
21. Giardine B, Riemer C, Hardison RC et al (2005) Galaxy: a platform for interactive large-scale genome analysis. Genome Res 5: 1451–1455
22. Conesa A, Madrigal P, Tarazona S et al (2016) A survey of best practices for RNA-seq data analysis. Genome Biol 17:13
23. Bolger AM, Lohse M, Usadel B (2014) Trimmomatic: a flexible trimmer for Illumina sequence data. Bioinformatics 30:2114–2120
24. Dobin A, Davis CA, Schlesinger F et al (2013) STAR: ultrafast universal RNA-seq aligner. Bioinformatics 29:15–21
25. Kim D, Paggi JM, Park C et al (2019) Graph-based genome alignment and genotyping with HISAT2 and HISAT-genotype. Nat Biotechnol 37:907–915
26. Li H (2018) Minimap2: pairwise alignment for nucleotide sequences. Bioinformatics 34: 3094–3100
27. Wu TD, Watanabe CK (2005) GMAP: a genomic mapping and alignment program for mRNA and EST sequences. Bioinformatics 21:1859–1875
28. Besemer J, Lomsadze A, Borodovsky M (2001) GeneMarkS: a self-training method for prediction of gene starts in microbial genomes. Implications for finding sequence motifs in regulatory regions. Nucleic Acids Res 29: 2607–2618
29. Stanke M, Morgenstern B (2005) AUGUSTUS: a web server for gene prediction in eukaryotes that allows user-defined constraints. Nucleic Acids Res 33:W465–W467
30. Kovaka S, Zimin AV, Pertea GM et al (2019) Transcriptome assembly from long-read RNA-seq alignments with StringTie2. Genome Biol 20:278
31. Trapnell C, Roberts A, Goff L et al (2012) Differential gene and transcript expression analysis of RNA-seq experiments with TopHat and Cufflinks. Nat Protocols 7:562–578
32. Haas BJ, Papanicolaou A, Yassour M et al (2013) De novo transcript sequence reconstruction from RNA-seq using the Trinity platform for reference generation and analysis. Nat Protocols 8:1494–1512
33. Keilwagen J, Hartung F, Grau J (2019) GeMoMa: homology-based gene prediction utilizing intron position conservation and RNA-seq data. Methods Mol Biol (Clifton, NJ) 1962:161–177
34. Hunter S, Apweiler R, Attwood TK et al (2008) InterPro: the integrative protein signature database. Nucleic Acids Res 37:D211–D215
35. Yao Z, You FM, N'Diaye A et al (2020) Evaluation of variant calling tools for large plant genome re-sequencing. BMC Bioinform 21: 360
36. Cingolani P, Platts A, Wang LL et al (2012) A program for annotating and predicting the effects of single nucleotide polymorphisms, SnpEff. Fly 6:80–92
37. Haile JK, N'Diaye A, Walkowiak S et al (2019) Fusarium head blight in durum wheat: recent status, breeding directions, and future research prospects. Phytopathology 109:1664–1675
38. Jin M, Liu H, He C et al (2016) Maize pan-transcriptome provides novel insights into

genome complexity and quantitative trait variation. Sci Rep 6:18936

39. Jupe F, Witek K, Verweij W et al (2013) Resistance gene enrichment sequencing (RenSeq) enables reannotation of the NB-LRR gene family from sequenced plant genomes and rapid mapping of resistance loci in segregating populations. Plant J 76:530–544
40. Liao Y, Smyth GK, Shi W (2013) featureCounts: an efficient general purpose program for assigning sequence reads to genomic features. Bioinformatics 30:923–930
41. Anders S, Pyl PT, Huber W (2015) HTSeq – a Python framework to work with high-throughput sequencing data. Bioinformatics 31:166–169
42. Lawrence M, Huber W, Pagès H et al (2013) Software for computing and annotating genomic ranges. PLoS Comput Biol 9:e1003118
43. Robinson MD, McCarthy DJ, Smyth GK (2010) edgeR: a bioconductor package for differential expression analysis of digital gene expression data. Bioinformatics 26:139–140
44. Love MI, Huber W, Anders S (2014) Moderated estimation of fold change and dispersion for RNA-seq data with DESeq2. Genome Biol 15:550
45. Shostak K, Bonner C, Sproule A et al (2020) Activation of biosynthetic gene clusters by the global transcriptional regulator *TRI6* in *Fusarium graminearum*. Mol Microbiol 114:664–680
46. Fauteux F, Wang Y, Rocheleau H et al (2019) Characterization of QTL and eQTL controlling early *Fusarium graminearum* infection and deoxynivalenol levels in a Wuhan 1 x Nyubai doubled haploid wheat population. BMC Plant Biol 19:536
47. Bolouri H (2014) Modeling genomic regulatory networks with big data. Trends Genet 30:182–191
48. Yu H, Huang J, Zhang W, Han J-DJ (2011) Network analysis to interpret complex phenotypes. Applied Statistics for Network Biology, pp 1–12
49. Langfelder P, Horvath S (2008) WGCNA: an R package for weighted correlation network analysis. BMC Bioinform 9:559
50. Zhang J, Huang K (2014) Normalized ImQCM: an algorithm for detecting weak quasi-cliques in weighted graph with applications in gene co-expression module discovery in cancers. Cancer Inform 13s3:CIN.S14021
51. Grimes T, Potter SS, Datta S (2019) Integrating gene regulatory pathways into differential network analysis of gene expression data. Sci Rep 9:5479
52. Shannon P, Markiel A, Ozier O et al (2003) Cytoscape: a software environment for integrated models of biomolecular interaction networks. Genome Res 13:2498–2504
53. Zhang M, Li Q, Yu D et al (2019) GeNeCK: a web server for gene network construction and visualization. BMC Bioinform 20:12
54. Jeong H, Mason SP, Barabási AL, Oltvai ZN (2001) Lethality and centrality in protein networks. Nature 411:41–42
55. Barabási A-L, Oltvai ZN (2004) Network biology: understanding the cell's functional organization. Nat Rev Genet 5:101–113
56. Farahbod M (2019) The interpretation of gene coexpression in systems biology. Doctoral dissertation, University of British Columbia. https://doi.org/10.14288/1.0387518
57. Li H, Sun Y, Zhan M (2009) Exploring pathways from gene co-expression to network dynamics. Methods Mol Biol (Clifton, NJ) 541:249–267
58. Sari E, Cabral AL, Polley B et al (2019) Weighted gene co-expression network analysis unveils gene networks associated with the Fusarium head blight resistance in tetraploid wheat. BMC Genomics 20:925
59. Gardiner DM, Kazan K, Manners JM (2009) Novel genes of Fusarium graminearum that negatively regulate deoxynivalenol production and virulence. Mol Plant-Microbe Interact 22: 1588–1600

Chapter 10

RNA-Seq Data Processing in Plant-Pathogen Interaction System: A Case Study

Ziying Liu, Youlian Pan, Yifeng Li, Thérèse Ouellet, and Nora A. Foroud

Abstract

In RNA-seq data processing, short reads are usually aligned from one species against its own genome sequence; however, in plant-pathogen interaction systems, reads from both host and pathogen samples are blended together. In contrast with single-genome analyses, both pathogen and host reference genomes are involved in the alignment process. In such circumstances, the order in which the alignment is carried out, whether the host or pathogen is aligned first, or if both genomes are aligned simultaneously, influences the read counts of certain genes. This is a problem, especially at advanced infection stages. It is crucial to have an appropriate strategy for aligning the reads to their respective genomes, yet the existing strategies of either sequential or parallel alignment become problematic when mapping mixed reads to their corresponding reference genomes. The challenge lies in the determination of which reads belong to which species, especially when homology exists between the host and pathogen genomes. This chapter proposes a combo-genome alignment strategy, which was compared with existing alignment scenarios. Simulation results demonstrated that the degree of discrepancy in the results is correlated with phylogenetic distance of the two species in the mixture which was attributable to the extent of homology between the two genomes involved. This correlation was also found in the analysis using two real RNA-seq datasets of *Fusarium*-challenged wheat plants. Comparisons of the three RNA-seq processing strategies on three simulation datasets and two real *Fusarium*-infected wheat datasets showed that an alignment to a combo-genome, consisting of both host and pathogen genomes, improves mapping quality as compared to sequential alignment procedures.

Key words RNA-seq, Host-pathogen interactions, Plant-microbe interaction system, Combo-genome, Sequential alignment, Parallel alignment, Bioinformatics

1 Introduction

High-throughput RNA-sequencing (RNA-seq) technology has become progressively accessible for functional genomics research and carries increased sensitivity and specificity compared to microarray technology [1]. A crucial task performed in RNA-seq data processing is aligning millions of short cDNA fragments (reads) to a reference genome. The percentage of uniquely mapped reads, a global indicator of the overall sequencing accuracy, is an important

Nora A. Foroud and Jonathan A. D. Neilson (eds.), *Plant-Pathogen Interactions*, Methods in Molecular Biology, vol. 2659,
https://doi.org/10.1007/978-1-0716-3159-1_10,

mapping quality measure [2]. Since alignment results will affect subsequent data analysis, it is important to select an unbiased sequence alignment procedure.

In the RNA-seq data processing for a single species, RNA-seq reads are aligned to their reference genome directly. In a study on plant-pathogen interaction, however, two genomes are involved; RNA-seq reads are thus a mixture from both pathogen and host genomes. Accurately determining whether the reads align to the host, the pathogen, or both genomes is challenging when separately performing alignment to a single genome at a time.

The RNA-seq technique enables researchers to study and monitor transcriptomic dynamics of host-pathogen interaction simultaneously under given conditions and furthermore examine the strategies applied by the host and pathogen to compete for limited biochemical resources [3, 4]. For example, Nuss et al. [5] developed an experimental approach that allows simultaneous monitoring of genome-wide infection-linked transcriptional alterations of the host and pathogen. They first split the libraries into reads originating from mouse and pathogen (*Yersinia pseudotuberculosis* IP32953) using Bowtie and Tophat2 and then excluded the identified cross-mapped reads from downstream analysis. Finally, the split RNA-seq reads were aligned to their corresponding genome respectively. Westermann's lab [6] applied dual-genome RNA-seq data of *Salmonella*-infected human cells to monitor the host response to bacterial infection. The mixture of RNA-seq reads originating from either *Salmonella enterica* or hosts (HeLa cells from human or mouse) was mapped to their respective genomes with high stringency. Mapped reads, which could be aligned equally well to both host and *Salmonella* reference sequences, were discarded in subsequent analysis. Bradford and colleagues [7] developed a species-specific mapping workflow for RNA-seq data from xenograft. The RNA-seq data were aligned to the human and mouse genomes separately. To differentiate human and mouse expression, the reads that mapped to both human and mouse genomes were removed, which resulted in loss of yields in each species (2–2.5% for human and 17–26% for mouse). Damron et al. [3] used dual RNA-seq to simultaneously measure *Pseudomonas aeruginosa* and the murine host's gene expression in response to respiratory infection. Reads data were mapped to *P. aeruginosa* and *Mus musculus* genomes independently by using the CLC Genomics platform. Dual-genome RNA-seq approaches were also applied to monitor host-viral interactions [8–11], assisting greatly in deciphering virus-driven disease mechanisms. Callari and colleagues [12] developed a method using in silico combined human-mouse reference genome for alignment to discriminate between human and mouse reads in patient-derived tumor xenograft to reduce the bias in the alignment.

In recent investigations into the transcriptomics associated with resistance and susceptibility of wheat against *Fusarium* head blight (FHB), the RNA-seq data contain reads from both wheat and the pathogen, *Fusarium graminearum* [13–16]. In this chapter, we present a case study wherein a combo-genome alignment approach is employed that maps RNA-seq reads to a combination of both host and pathogen genomes. This approach was compared with two scenarios processing RNA-seq read alignments sequentially using three simulation datasets and two real-life datasets on *Fusarium*-affected wheat.

2 Materials

2.1 Datasets

1. Simulated datasets: Genomes of *Arabidopsis thaliana* and three other species were selected to simulate RNA-seq datasets for this case study, with differences in their phylogenetic distances consisting of two dicots, *Arabidopsis lyrata* in the same genus, *Brassica rapa* in the same family, and one monocot *Brachypodium distachyon* (Fig. 1a).
2. *Fusarium*-affected dataset GSE118126: Dataset GSE118126 [14] from the Gene Expression Omnibus of the National Center for Biotechnology Information (NCBI-GEO) repository was used in our study. Each of the four wheat genotypes, including FHB-resistant Sumai3 and FL62R1 and FHB-susceptible Stettler and Muchmore, were inoculated with *F. graminrearum*. Samples were collected at 0, 1, 2, and 3 days post-inoculation (dpi). There were three biological replicates for each sample.
3. *Fusarium*-affected dataset GSE113128: Dataset GSE11312 8 [13] from NCBI-GEO was also used in our experiments. Each of the four wheat genotypes, including FHB-resistant Wuhan1, Nyubai and HC374, and FHB-susceptible Shaw, were inoculated with *F. graminrearum*. Samples were collected at 2 and 4 dpi. There were three biological replicates for each sample.

2.2 Genome References and Corresponding Annotation Data

1. The genome references and corresponding annotation files of *A. thaliana*, *A. lyrata*, *B. rapa*, and *B. distachyon* (Fig. 1a) were obtained from EnsemblPlants (release-52) [17].
2. Reference genomes for *Fusarium* and wheat. The wheat IWGSC refseq version 2.1 genome assembly and the corresponding gene annotation were obtained from the IWGSC website [18]. The *F. graminearum* genome (*F. graminearum* str. PH-1 [RR1]) and the corresponding gene annotation files were obtained from EnsemblFungi (release-52) [19].

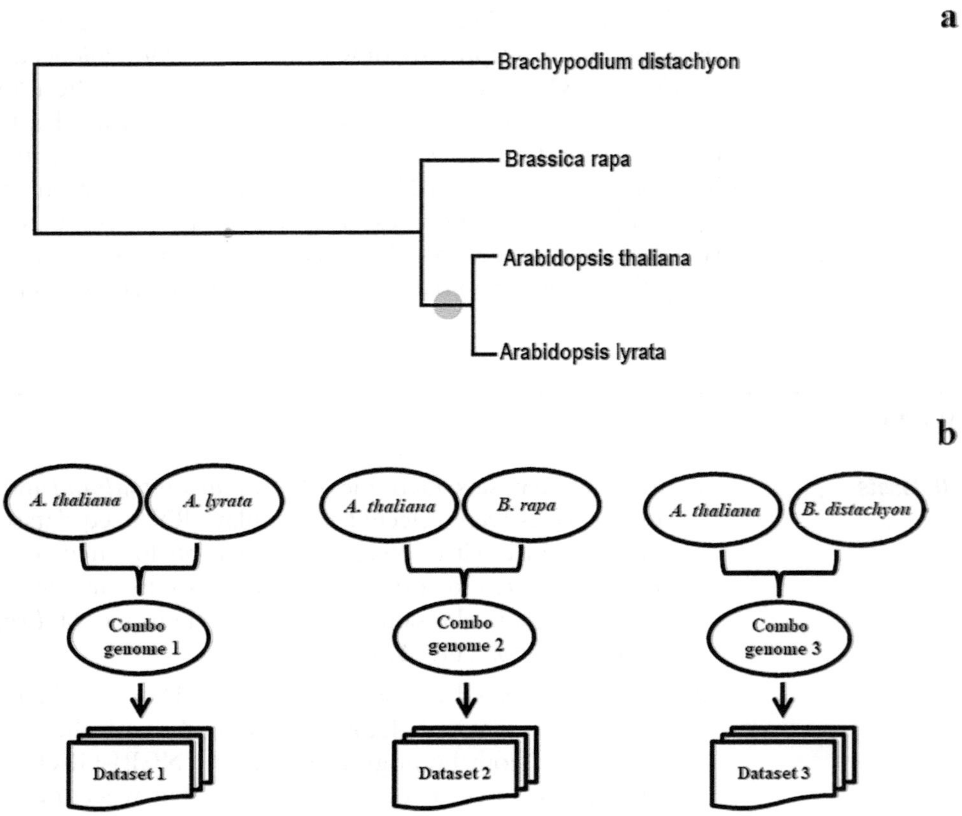

Fig. 1 Four plant species involved in the simulation. (**a**) Phylogenetic distances between *A. thaliana* and each of the three other species. Phylogenetic tree was created with iTOL [23]. (**b**) The simulation process to generate dual-genome RNA-seq datasets

2.3 Computer Hardware

1. Hard drives 6 TB.
2. Memory 128GB. Minimum requirement 32 GB.
3. Processor Intel(R) Xeon(R) Gold 5118 CPU at 2.30 GHz and 2.29 GHz (two processors). In general, high-speed computer with more memory will process RNA-seq data faster.

2.4 Software

1. Ubuntu 18.04.6 LTS.
2. Flux-simulator [20]. Under Linux System using conda to install flux-simulator:

```
$conda install -c bioconda flux-simulator
```

3. Precompiled binaries FASTX version 0.0.13 [21] for Linux (64bit). It can be downloaded from the following website http://hannonlab.cshl.edu/fastx_toolkit/download.html (*see* **Note 1**).
4. Binaries version of STAR [22]. It can be downloaded from GitHub: https://github.com/alexdobin/STAR (*see* **Note 2**).

5. iTOL [23] using the web server at https://itol.embl.de/.
6. Diamond [24] can be downloaded from https://github.com/bbuchfink/diamond (*see* **Note 3**).

3 Method

In a dual-genome system, such as the host-pathogen system, two reference genomes and their corresponding annotations were combined into a combo-reference genome and a combo-annotation, respectively. RNA-seq reads were aligned to the combo-reference. This approach was compared with two scenarios of sequential alignment using three simulated datasets and two real-life datasets on *Fusarium*-infected wheat.

3.1 Mapping Strategies on Simulated Datasets

The simulated datasets were used to demonstrate the impact of homology on the mapping quality of a dual-genome RNA-seq alignment.

3.1.1 Generate Simulation Data

1. Download the reference genomes and annotation files of *A. thaliana*, *A. lyrata*, *B. rapa*, and *B. distachyon* (Fig. 1a) from the EnsemblPlants (Release 52) [17].

```
Arabidopsis_thaliana.TAIR10.dna_sm.toplevel.fa
Arabidopsis_thaliana.TAIR10.52.gtf

Arabidopsis_lyrata.v.1.0.dna_sm.toplevel.fa
Arabidopsis_lyrata.v.1.0.52.gtf
Brassica_rapa.Brapa_1.0.dna_sm.toplevel.fa
Brassica_rapa.Brapa_1.0.52.gtf
Brachypodium_distachyon_v3.0.dna_sm.toplevel.fa
Brachypodium_distachyon_v3.0.52.gtf
```

2. Combine *A. thaliana* with each of the other three genomes, respectively:

```
$cat Arabidopsis_lyrata.v.1.0.52.gtf Arabidop-
sis_thaliana.TAIR10.52.gtf > combinedAnnotation_-
lyrata_thaliana.gtf

$cat Arabidopsis_lyrata.v.1.0.dna_sm.toplevel.fa
Arabidopsis_thaliana.TAIR10.dna_sm.toplevel.fa >
combinedGenome_lyrata_thaliana.fa

$cat Brassica_rapa.Brapa_1.0.52.gtf Arabidop-
sis_thaliana.TAIR10.52.gtf > combinedAnnotation_-
Brapa_thaliana.gtf
```

```
$cat Brassica_rapa.Brapa_1.0.dna_sm.toplevel.fa
 Arabidopsis_thaliana.TAIR10.dna_sm.toplevel.fa >
combinedGenome_Brapa_thaliana.fa

 $cat Brachypodium_distachyon_v3.0.52.gtf Arabi-
dopsis_thaliana.TAIR10.52.gtf > combinedAnnota-
tion_Bdistachyon_thaliana.gtf

 $cat Brachypodium_distachyon_v3.0.dna_sm.tople-
vel.fa
 Arabidopsis_thaliana.TAIR10.dna_sm.toplevel.fa >
combinedGenome_Bdistachyon_thaliana.fa
```

3. Each of the three genome combinations was used to generate 50 million paired-end synthetic RNA-seq datasets, 101 bps in length, by using the Flux-Simulator (version 1.2.1) with default error model [23] (Fig. 1b) (*see* **Note 4**).

```
$flux-simulator -p parameters.par
```

4. `r1.fastq` and `r2.fastq` files were simulated RNA-seq data for each combo-genome generated from **step 2** (Fig. 1b).

3.1.2 Strategy A: Perform Dual-Genome Alignment Using Simulated Datasets

Align RNA-seq reads (`r1.fastq` and `r2.fastq`) to the combined genome with combined annotation (Fig. 2a).

1. Combine genome 1 and genome 2 reference genomes (.fa) and annotation files (.gtf)

```
$cat genome1Annotation.gtf genome2Annotation.gtf >
combinedAnnotation.gtf
$cat genomeRef1.fa genomeRef2.fa > combinedGeno-
meRef.fa
```

2. Build STAR index using combined genome reference:

```
$STAR --runThreadN ${num_thread} \
--runMode genomeGenerate \
--genomeDir ${combined_ref_index_dir}\
--genomeFastaFiles ${combinedGenomeRef.fa}\
--sjdbGTFfile ${combinedAnnotation.gtf} \
--sjdbOverhang ${readlength -1} \
--limitGenomeGenerateRAM 100000000000
```

3. Perform alignment

```
$STAR --runMode alignReads \
--runThreadN ${num_thread} \
--limitBAMsortRAM 100000000000 \
--limitSjdbInsertNsj 5000000 \
```

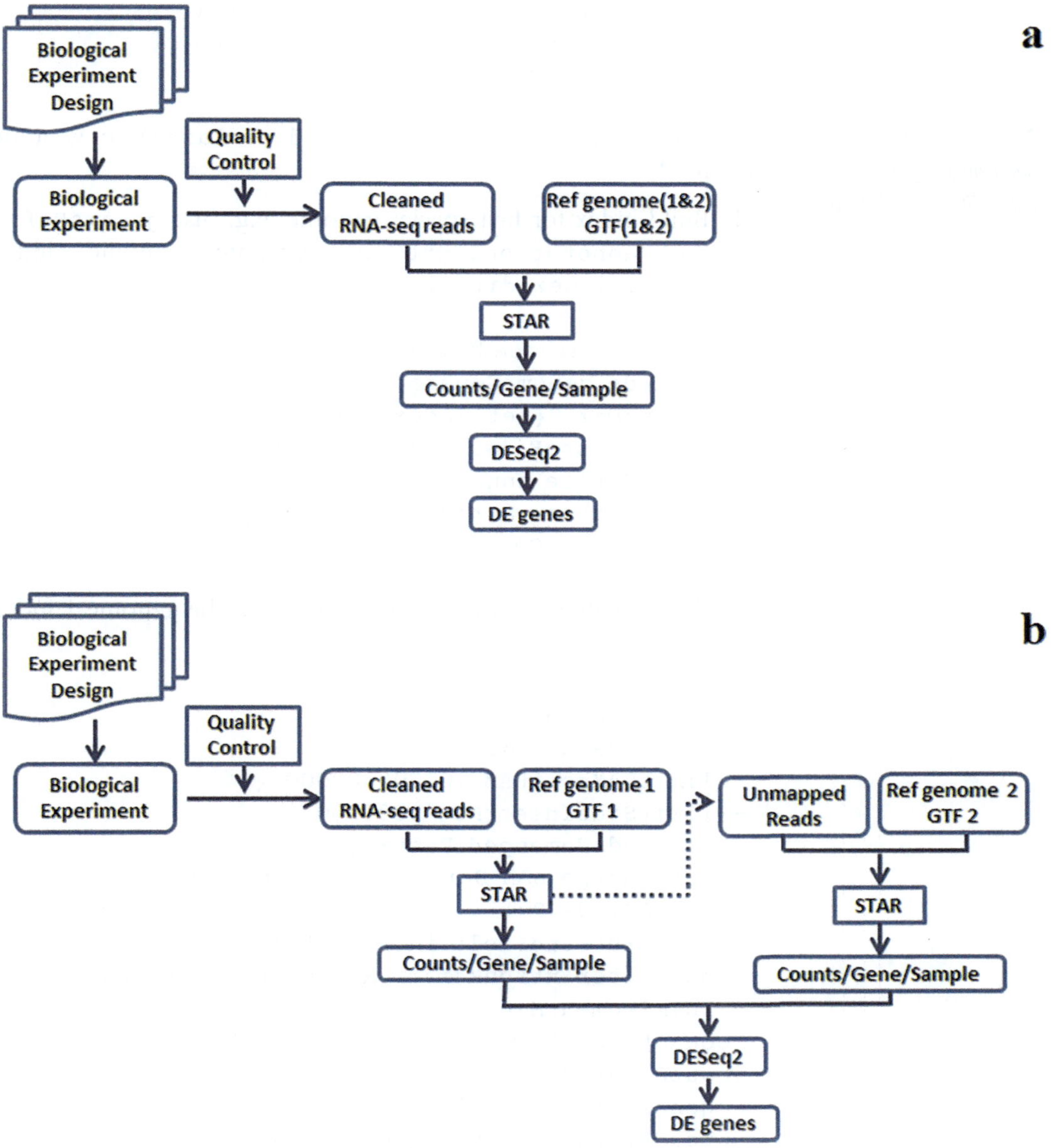

Fig. 2 RNA-seq data analysis pipelines: (**a**) Dual-genome alignment workflow and (**b**) sequential alignment workflow

```
--outReadsUnmapped Fastx \
--outSAMtype BAM SortedByCoordinate \
--outSAMmode Full \
--outSAMstrandField intronMotif \
--outFilterIntronMotifs RemoveNoncanonical \
--chimSegmentMin 20 \
--quantMode TranscriptomeSAM GeneCounts \
--outSAMattributes All \
```

```
--genomeDir ${combined_ref_index_dir} \
--readFilesIn ${r1.fastq} ${r2.fastq} \
--outFileNamePrefix ${output_dir}
```

3.1.3 Strategy B: Perform Sequential Alignments Using Simulated Datasets

Align RNA-seq reads (`r1.fastq` and `r2.fastq`) to genome 1 (Fig. 2b).

1. Build index for first species genome, e.g., use `genomeRef1.fa`, `annotation1.gtf` to generate genome index 1 `{ref_index_dir1}`:

```
$STAR --runThreadN ${num_thread} \
--runMode genomeGenerate \
--genomeDir ${ref_index_dir1} \
--genomeFastaFiles genomeRef1.fa \
--sjdbGTFfile annotation1.gtf \
--sjdbOverhang ${readlength -1} \
--limitGenomeGenerateRAM 100000000000
```

2. Perform alignment of the mixed reads to the genome 1 index `{ref_index_dir1}`:

```
$STAR --runMode alignReads \
--runThreadN ${num_thread} \
--limitBAMsortRAM 100000000000 \
--limitSjdbInsertNsj 5000000 \
--outReadsUnmapped Fastx \
--outSAMtype BAM SortedByCoordinate \
--outSAMmode Full \
--outSAMstrandField intronMotif \
--outFilterIntronMotifs RemoveNoncanonical \
--chimSegmentMin 20 \
--quantMode TranscriptomeSAM GeneCounts \
--outSAMattributes All \
--genomeDir ${ref_index_dir1} \
--readFilesIn r1.fastq r2.fastq \
--outFileNamePrefix ${output_dir}
```

3. Build index `{ref_index_dir2}` for the second species genome using genomeRef2.fa, annotation2.gtf

```
$STAR --runThreadN ${num_thread} \
--runMode genomeGenerate \
--genomeDir ${ref_index_dir2} \
--genomeFastaFiles genomeRef2.fa \
--sjdbGTFfile annotation2.gtf \
--sjdbOverhang ${readlength -1} \
--limitGenomeGenerateRAM 100000000000
```

4. Perform alignment of the unmapped mixed reads from **Step 2** (`r1_unmapped.fastq, r2_unmapped.fastq`) to the genome 2 index `{ref_index_dir2}`

```
$STAR --runMode alignReads \
--runThreadN ${num_thread} \
--limitBAMsortRAM 100000000000 \
--limitSjdbInsertNsj 5000000 \
--outReadsUnmapped Fastx \
--outSAMtype BAM SortedByCoordinate \
--outSAMmode Full \
--outSAMstrandField intronMotif \
--outFilterIntronMotifs RemoveNoncanonical \
--chimSegmentMin 20 \
--quantMode TranscriptomeSAM GeneCounts \
--outSAMattributes All \
--genomeDir ${ref_index_dir2} \
--readFilesIn r1_unmapped.fastq r2_unmapped.fastq
\
--outFileNamePrefix ${output_dir}
```

3.1.4 Strategy C: Perform Sequential Alignment in Reverse Order of Strategy B Using Simulated Datasets

1. Perform alignment of the mixed reads to genome 2 index `{ref_index_dir2}` generated from Subheading 3.1.3, **step 3** (Fig. 2b):

```
$STAR --runMode alignReads \
--runThreadN ${num_thread} \
--limitBAMsortRAM 100000000000 \
--limitSjdbInsertNsj 5000000 \
--outReadsUnmapped Fastx \
--outSAMtype BAM SortedByCoordinate \
--outSAMmode Full \
--outSAMstrandField intronMotif \
--outFilterIntronMotifs RemoveNoncanonical \
--chimSegmentMin 20 \
--quantMode TranscriptomeSAM GeneCounts \
--outSAMattributes All \
--genomeDir ${ref_index_dir2} \
--readFilesIn r1.fastq r2.fastq \
--outFileNamePrefix ${output_dir}
```

2. Align the unmapped reads (`r1_unmapped.fastq, r2_unmapped.fastq`) in `{output_dir}` to genome index 1 `{ref_index_dir1}`, which was generated from Subheading 3.1.3, **Step 1**

```
$STAR --runMode alignReads \
--runThreadN ${num_thread} \
--limitBAMsortRAM 100000000000 \
--limitSjdbInsertNsj 5000000 \
--outReadsUnmapped Fastx \
--outSAMtype BAM SortedByCoordinate \
--outSAMmode Full \
--outSAMstrandField intronMotif \
--outFilterIntronMotifs RemoveNoncanonical \
--chimSegmentMin 20 \
--quantMode TranscriptomeSAM GeneCounts \
--outSAMattributes All \
--genomeDir ${ref_index_dir1} \
--readFilesIn r1_unmapped.fastq r2_unmapped.fastq
\
--outFileNamePrefix ${output_dir}
```

In the simulation results, the mapping rate of sequential alignment varied significantly for the pairs of *A. thaliana* and *A. lyrata*, depending on the mapping order, showing a difference of 49.3–50.32%. This gap narrowed for more distant pairs between *A. thaliana* and *B. rapa* (8.21–8.31%) and minimal between the most distant pair *A. thaliana* and *B. discachyon* (0.01–0.03%) (Table 1).

The difference in mapping rates between the two sequential alignments was significantly correlated with the extent of homology between the two genomes ($r = 0.88$, $p < 0.05$). The overall mapping rates of the combo-genome alignment were similar or equal to the sum of sequential alignments.

The difference of the unique mapping rate becomes less significant when there are fewer homologs between the genomes such as in the pairs of *A. thaliana* and *B. distachyon*. When mapping to the combo-genome, various possible alignments are considered and the optimal one is usually picked, i.e., a short read will be aligned to the most matched loci. For alignment of the similar quality reads, the mapping to both references equally well will be reported as multi-mapping reads [22]. This can help reduce the false positives and false negatives for unique calls to each species when compared with sequential alignment. In this way, less information will be lost.

3.2 Mapping Strategies for the F. graminearum-Wheat Interaction Datasets

In the case of a pathogen-host system, the alignment of the reads to the actual genome is difficult because it is not practical to completely isolate the host RNA-seq reads from the pathogen, due to the presence of homologous genes in the two genomes. In a sequential alignment, the mapping result is strongly biased toward the first reference genome in the processing order (Table 1). The extent of the bias depends on the extent of

Table 1
Simulation results

	Strategy	Genome	Unique alignment (%)
Simulation Dataset 1	A	Combo-genome	81.68
	B	*A. thaliana*	74.30
		A. lyrata	7.9
	C	*A. lyrata*	57.20
		A. thaliana	23.98
Simulation Dataset 2	A	Combo-genome	91.32
	B	*A. thaliana*	61.94
		B. rapa	29.30
	C	*B. rapa*	37.51
		A. thaliana	53.63
Simulation Dataset 3	A	Combo-genome	92.68
	B	*A. thaliana*	40.91
		B. distachyon	51.75
	C	*B. distachyon*	51.78
		A. thaliana	40.90

homology between the two genomes. As a consequence, this bias will translate to errors in the subsequent data quantification, normalization, and differentially expressed gene identification. The purpose of the combo-genome strategy introduced in this chapter is to mitigate such bias, and we applied this strategy using real datasets (GSE118126 and GSE113128) as an example.

3.2.1 Preprocess RNA-Seq Reads

1. Load raw fastq files (`r1.fastq` and `r2.fastq`) to the script `fastq_quality_trimmer` in the FASTX-Toolkit package [20] to remove low-quality reads for each individual fastq file:

```
$fastq_quality_trimmer -Q 33 -v -t 20 -l 20 -i r1.
fastq -o r1_trimmed.fastq
$fastq_quality_trimmer -Q 33 -v -t 20 -l 20 -i r2.
fastq -o r2_trimmed.fastq
```

2. Load `r1_trimmed.fastq` and `r2_trimmed.fastq` and adaptor sequence to the script `fastx-clipper` in FASTX-Toolkit package [20] to remove adaptor sequences for each individual fastq file:

```
$fastx_clipper -Q 33 -a adaptorSeq -l 20 -n -v -i
r1_trimmed.fastq -o r1_trimmedClipped.fastq
$fastx_clipper -Q 33 -a adaptorSeq -l 20 -n -v -i
r2_trimmed.fastq -o r2_trimmedClipped.fastq
```

3. Re-pair `r1_trimmedClipped.fastq` and `r2_trimmedClipped.fastq` using the `fastqcombinepairedend.py` script in the FASTX-Toolkit package [20] (*see* **Note 5**)

```
$fastqcombinepairedend.py "@NB5" " " r1_trimmed-
Clipped.fastq r2_trimmedClipped.fastq
```

The output files `r1_trimmedClipped_stillPaired.fastq` and `r2_trimmedClipped_stillPaired.fastq` will be generated and used in the subsequent alignment.

3.2.2 Obtain and Combine Wheat and F. graminearum Reference Genome and Annotation Files

1. Download the wheat reference genome and its corresponding annotation from the IWGSC website [18], and the *F. graminearum* reference genome assembly and its corresponding annotation from the EnsemblFungi website [19]:

```
iwgsc_refseqv2.1_assembly.fa
iwgsc_refseqv2.1_annotation_200916_HC.gtf
Fusarium_graminearum.RR1.dna_sm.toplevel.fa
Fusarium_graminearum.RR1.52.gtf
```

2. Combine wheat and *F. graminearum* reference genomes and their annotations, respectively:

```
$cat iwgsc_refseqv2.1_annotation_200916_HC.gtf,
Fusarium_graminearum.RR1.52.gtf > combinedAnnota-
tion.gtf
$cat iwgsc_refseqv2.1_assembly.fa Fusarium_grami-
nearum.RR1.dna_sm.toplevel.fa > combinedGenomeRef.
fa
```

3. Build the STAR index {`combined_ref_index_dir`}

```
$STAR --runThreadN ${num_thread} \
--runMode genomeGenerate \
--genomeDir ${combined_ref_index_dir} \
--genomeFastaFiles combinedGenomeRef.fa \
--sjdbGTFfile ${combinedAnnotation.gtf} \
--sjdbOverhang 100 \
--limitGenomeGenerateRAM 100000000000
```

4. Perform alignment by aligning mixed reads to the index generated using two genomes.

```
$STAR --runMode alignReads \
--runThreadN ${num_thread} \
```

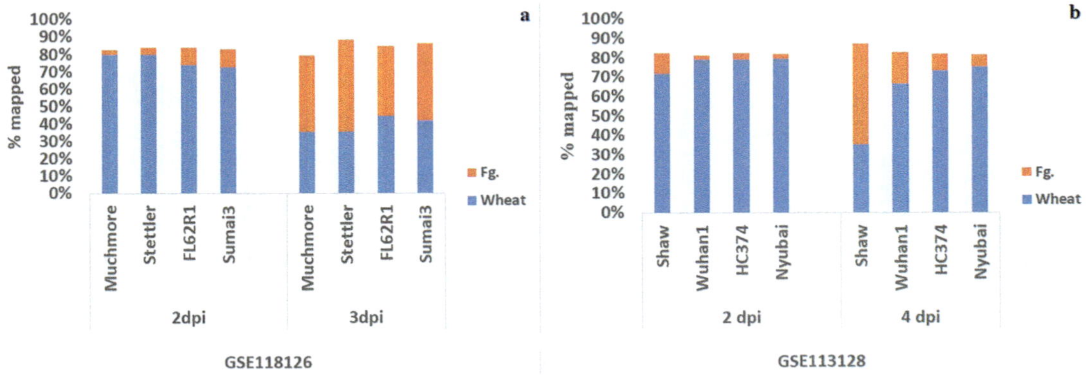

Fig. 3 Dynamic interactome between wheat and *F. graminearum* (Fg.) using results from strategy A. (**a**) *F. graminearum* inoculation from susceptible to resistance genotypes at 2 and 3 dpi of GSE118126 and (**b**) from susceptible (Shaw) to resistance genotypes at 2 and 4 dpi of GSE113128

```
 --limitBAMsortRAM 100000000000 \
 --limitSjdbInsertNsj 5000000 \
 --outReadsUnmapped Fastx \
 --outSAMtype BAM SortedByCoordinate \
 --outSAMmode Full \
 --outSAMstrandField intronMotif \
 --outFilterIntronMotifs RemoveNoncanonical \
 --chimSegmentMin 20 \
 --quantMode TranscriptomeSAM GeneCounts \
 --outSAMattributes All \
 --genomeDir ${combined_ref_index_dir} \
 --readFilesIn r1_trimmedClipped_stillPaired.fastq
r2_trimmedClipped_stillPaired.fastq \
 --outFileNamePrefix ${output_dir}
```

On the two real datasets, the number of pathogen reads increased over time reaching a peak at 3 dpi (52.82%) and 4 dpi (51.94%) for GSE118126 and GSE113128, respectively (Fig. 3a, 3b). The number of pathogen reads in resistance genotypes of GSE113128 was much lower than that in susceptible genotypes (Fig. 3b). The mapping results varied in the sequential alignment depending on the alignment order—host first (strategy B) or pathogen first (strategy C)—and consistent with simulation results. The differences among the mapping rates are presented in Fig. 4.

3.3 Technical Insights

In this case study, we demonstrated a strategy of aligning the RNA-seq reads to the combo-genome references and compared its results with sequential alignment strategies using simulated datasets and real datasets. In the case of a pathogen-host system, the alignment of the reads to its own genome is difficult because it is not practical

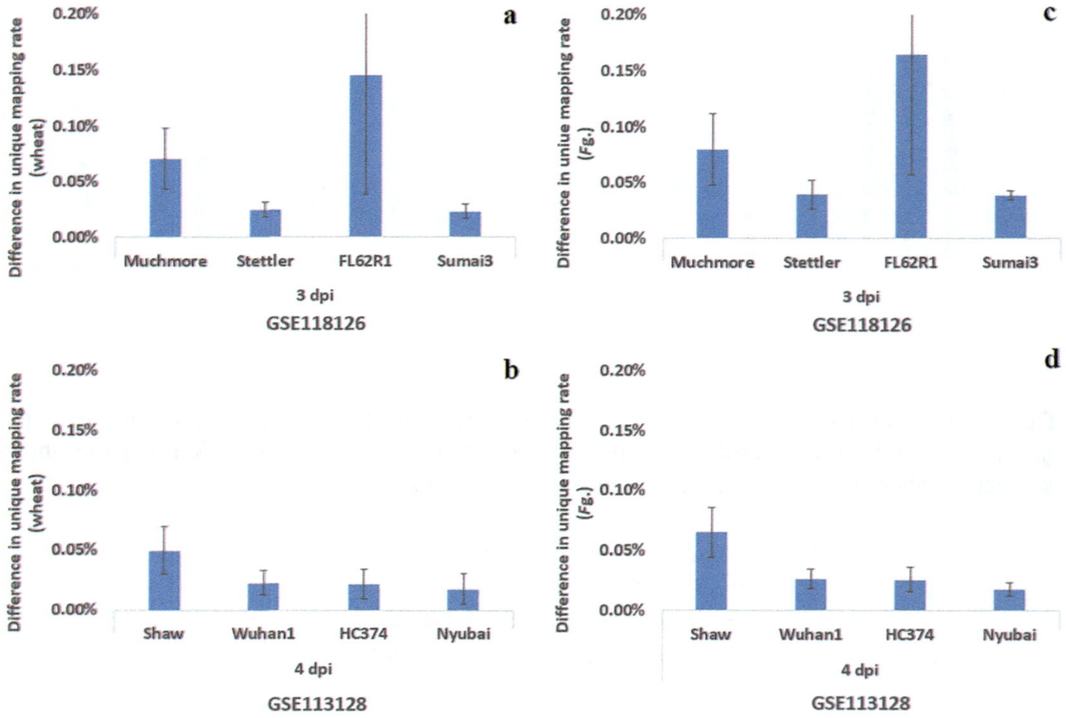

Fig. 4 Differences in unique mapping rates between two sequential alignment strategies; (**a**) and (**b**) are the differences in wheat unique mapping between strategies B and C for GSE118126 and GSE1113128, respectively, while (**c**) and (**d**) are differences in *F. graminearum* unique mapping. Error bar indicates standard error of the mean

to isolate the host RNA-seq reads from the pathogen completely, and homologous genes exist between the two genomes. In a sequential alignment, the mapping result is strongly biased toward the first reference genome in the processing order. The extent of the bias depends on the proportion of homology between the two genomes. This strategy reaches a higher accuracy and reduces the number of false positives and further reduces the bias in the subsequent data analysis. The intention of the combo-genome strategy introduced in this case study is to mitigate such bias.

4 Notes

1. The pre-compiled binary version of the FASTX-toolkit is easy to use: The users have to choose the one matching their operating systems. In our experiments, we used Linux (64bit). After downloading the executable binaries, users should add the FASTX-Toolkit executable to the PATH variable, e.g., $ export PATH = $PATH:/path/to/fastx_bin/bin.

Alternatively, cutadapt or scythe can be used to remove adaptor sequences and low-quality reads.

2. The pre-compiled STAR executables are located in bin/subdirectory. The static executables are the easiest to use, as they are statically compiled and are not dependents on external libraries [22]. Users need to add the STAR executable to the PATH variable, e.g., $ export PATH = $PATH:/path/to/STAR-master_2.7.9a/bin/Linux_x86_64_static/STAR.
3. Software Diamond usage can be found at https://github.com/bbuchfink/diamond/wiki. Users may download and install the software as follows:

```
$wget http://github.com/bbuchfink/diamond/re-
leases/download/v2.0.13/diamond-linux64.tar.gz
# untar the compressed file
$tar xzf diamond-linux64.tar.gz

# set path
$ export PATH=$PATH:/path/to/diamond
```

4. Below is an example of parameters.par

```
REF_FILE_NAME combinedAnnotation_lyrata_thaliana.
gtf
GEN_DIR /combinedAnnotation_lyrata_thaliana
NB_MOLECULES 100000
READ_NUMBER 10000

# use default 76-bp error model
ERR_FILE 76

# create a fastq file
FASTA YES
PAIRED_END true

# Length of the reads
#
# number default: 36
READ_LENGTH 101

# Number of reads sequenced
#
# number default: 5000000
READ_NUMBER 50000000

# Switch on/off Reverse Transcription
```

```
#
# true|false or yes|no default: true
RTRANSCRIPTION true

# Mean value of a gaussian distribution that re-
flects GC bias
# amplification chance, set to 'NaN' to disable GC
biases. 0.0<=
# number <= 1.0 default: 0.5
GC_MEAN 0.34
```

We defined "REF_FILE_NAME" as "combinedAnnotation_-lyrata_thaliana.gtf", which is the combined annotation GTF file; "GEN_DIR" as "/combinedAnnotation_lyrata_thaliana", which is the folder where we put the genome references; "PAIRED_END" and "FASTA" as "true" and "YES", respectively, for generated paired-end fastq sequences; "READ_LENGTH" as "101" and "READ_NUMBER" as "50,000,000" to generate 50 million read sequences with read length 101. "combinedAnnotation_lyr-ata_thaliana.gtf" can be replaced by a different annotation file, and under "/combinedAnnotation_lyrata_thaliana" folder are the corresponding genome references.

5. To run this script, users first will need to find the information in the header line of the fastq file. For example, "`@NB501604:57:HJ32LAFXY:1:11101:16749:2515 1:N:0:ACTTGA`". The script in this step requires four parameters: (1) the first four characters of the header line, such as "`@NB5`" in this case; (2) the separator before r1 or r2 indicator, such as a space " " in this case; (3) `r1_*.fastq`; and (4) `r2_*.fastq` to be re-paired.

Acknowledgments

We are thankful to Dr. Linda Harris (Agriculture and Agri-Food Canada) and Dr. Sean Walkowiak (Canadian Grain Commission) for their insightful discussion at the early stage of this project.

References

1. Nagalakshmi U, Wang Z, Waern K et al (2008) The transcriptional landscape of the yeast genome defined by RNA sequencing. Science 320:1344–1349
2. Conesa A, Madrigal P, Tarazona S et al (2016) A survey of best practices for RNA-seq data analysis. Genome Biol 17:13
3. Damron FH, Oglesby-Sherrouse AG, Wilks A, Barbier M (2016) Dual-seq transcriptomics reveals the battle for iron during *Pseudomonas aeruginosa* acute murine pneumonia. Sci Rep 6:39172
4. Westermann AJ, Barquist L, Vogel J (2017) Resolving host-pathogen interactions by dual RNA-seq. PLoS Pathog 13:e1006033

5. Nuss AM, Beckstettea M, Pimenovaa M et al (2017) Tissue dual RNA-seq allows fast discovery of infection-specific functions and riboregulators shaping host-pathogen transcriptomes. Proc Natl Acad Sci 114: E791–E800
6. Westermann AJ, Gorski SA, Voge J (2012) Dual RNA-seq of pathogen and host. Nat Rev Microbiol 10:618–630
7. Bradford JR, Farren M, Powell SJ et al (2013) RNA-seq differentiates tumour and host mRNA expression changes induced by treatment of human tumour xenografts with the VEGFR tyrosine kinase inhibitor cediranib. PLoS One 8:e66003
8. Woodhouse SD, Narayan R, Latham S et al (2010) Transcriptome sequencing, microarray, and proteomic analyses reveal cellular and metabolic impact of hepatitis C virus infection *in vitro*. Hepatology. 52:443–453
9. Strong MJ, Xu G, Coco J et al (2013) Differences in gastric carcinoma microenvironment stratify according to EBV infection intensity: implications for possible immune adjuvant therapy. PLoS Pathog 9:e1003341
10. Perez-Losada M, Castro-Nallar E, Bendall ML et al (2015) Dual transcriptomic profiling of host and microbiota during health and disease in pediatric asthma. PLoS One. 10:e0131819
11. Wesolowska-Andersen A, Everman JL, Davidson R et al (2017) Dual RNA-seq reveals viral infections in asthmatic children without respiratory illness which are associated with changes in the airway transcriptome. Genome Biol 18: 12
12. Callari M, Batra AS, Batra RN et al (2018) Computational approach to discriminate human and mouse sequences in patient-derived tumour xenografts. BMC Genomics 19:19
13. Pan Y, Liu Z, Rocheleau H et al (2018) Transcriptome dynamics associated with resistance and susceptibility against fusarium head blight in four wheat genotypes. BMC Genomics 19: 642
14. Wang L, Li Q, Liu Z et al (2018) Integrated transcriptome and hormone profiling highlight the role of multiple phytohormone pathways in wheat resistance against fusarium head blight. PLoS One 13:e0207036
15. Fauteux F, Wang Y, Rocheleau H et al (2019) Characterization of QTL and eQTL linked with early *Fusarium graminearum* infection and deoxynivalenol levels in a Wuhan 1 x Nyubai doubled-haploid wheat population. BMC Plant Biol 19:1451
16. Buhrow LM, Liu Z, Cram D et al (2021) Wheat transcriptome profiling reveals abscisic and gibberellic acid treatments regulate early-stage phytohormone defense signaling, cell wall fortification, and metabolic switches following *Fusarium graminearum*-challenge. BMC Genomics 22:798
17. Howe KL, Achuthan P, Allen J et al (2021) Ensembl. Nucleic Acids Res 49:884–891
18. Zhu T, Wang L, Rimbert H et al (2021) Optical maps refine the bread wheat *Triticum aestivum* cv Chinese Spring genome assembly. Plant J 107:303–314
19. http://ftp.ensemblgenomes.org/pub/fungi/release-52/
20. Griebel T, Zacher B, Ribeca P et al (2012) Modelling and simulating generic RNA-seq experiments with the flux simulator. Nucleic Acids Res 40(10073–10):083
21. Hannon GJ (2010) FASTX-Toolkit. http://hannonlab.cshl.edu/fastx_toolkit
22. Dobin A, Davis CA, Schlesinger F et al (2013) STAR: ultrafast universal RNA-seq aligner. Bioinform 29:15–21
23. Letunic I, Bork P (2011) Interactive Tree Of Life v2: online annotation and display of phylogenetic trees made easy. Nucleic Acids Res. 39:W475–W478
24. Buchfink B, Reuter K, Drost HG (2021) Sensitive protein alignments at tree-of-life scale using DIAMOND. Nat Methods 18:366–368

Chapter 11

Differential Expression Feature Extraction (DEFE): A Case Study in Wheat FHB RNA-Seq Data Analysis

Youlian Pan, Anuradha Surendra, Ziying Liu, Thérèse Ouellet, and Nora A. Foroud

Abstract

In differential gene expression data analysis, one objective is to identify groups of co-expressed genes from a large dataset in order to detect the association between such a group of genes and an experimental condition. This is often done through a clustering approach, such as *k*-means or bipartition hierarchical clustering, based on particular similarity measures in the grouping process. In such a dataset, the gene differential expression itself is an innate attribute that can be used in the feature extraction process. For example, in a dataset consisting of multiple treatments versus their controls, the expression of a gene in each treatment would have three possible behaviors, upregulated, downregulated, or unchanged. We present in this chapter, a differential expression feature extraction (DEFE) method by using a string consisting of three numerical values at each character to denote such behavior, i.e., 1 = up, 2 = down, and 0 = unchanged, which results in up to 3^B differential expression patterns across all B comparisons. This approach has been successfully applied in many research projects, and among these, we demonstrate the strength of DEFE in a case study on RNA-sequencing (RNA-seq) data analysis of wheat challenged with the phytopathogenic fungus, *Fusarium graminearum*. Combinations of multiple schemes of DEFE patterns revealed groups of genes putatively associated with resistance or susceptibility to FHB.

Key words Differential gene expression, Differential expression feature extraction, RNA-seq, Co-expression, Gene regulation, Wheat, Fusarium head blight, *Fusarium graminearum*, Bioinformatics

1 Introduction

High-throughput sequencing technology revolutionized the landscape of biology. It produces short fragments of sequences at various depths, which are quantified as read counts, transcripts per million (TPM), fragments per kilobase of transcript per million fragments mapped (FPKM), and reads per kilobase of transcript per million fragments mapped (RPKM). Using tools such as DESeq2

Supplementary Information The online version contains supplementary material available at https://doi.org/10.1007/978-1-0716-3159-1_11.

Nora A. Foroud and Jonathan A. D. Neilson (eds.), *Plant-Pathogen Interactions*, Methods in Molecular Biology, vol. 2659, https://doi.org/10.1007/978-1-0716-3159-1_11,

[1], Cuffdiff [2], and edgeR [3], differential expression signals can be detected from the read counts rather than the three per million fragments normalized quantifications (TPM, FPKM, RPKM) [4, 5].

Co-expressed genes are identified based on similarity in their expression profiles over many treatments in an experiment. Gene co-expression analysis is usually done through various clustering methods based on similarity characteristics in the data. Association of a cluster of co-expressed genes with certain phenotypic traits is subsequently inferred. Clustering is an art and can generate beautiful as well as unattractive grouping results. Subjectivity is hard to avoid in clustering. For example, in *k*-means clustering, the definition of the *k* value is more or less subjective. As a result, multiple *k* values would have to be applied and this subjectivity can be reduced by various methods, such as Silhouette [6] and cluster stability [7], as applied in our earlier research in this field [8]. The final *k* value will still be more or less subjective based on the research objective and on the optimum number of patterns, which is usually unknown. The second issue in clustering is the selection of distance measure. For example, pairwise Pearson correlation is often used to retrieve groups based on similarity in variation patterns; Euclidean distance is often considered in grouping based on difference in overall values across the samples. In more complicated cases, subspace clustering is often considered [9].

In a set of differentially expressed genes, the pattern of differential expression over a series of experimental conditions of the genes themselves are innate attribute that can be used in the feature extraction process. In this chapter, we present a theoretically simple and practical method to find gene differential expression patterns from large RNA-seq datasets in order to extract features (genes) that are closely related with the objectives of a research project. The method is called differential expression feature extraction (DEFE) and outperforms two conventional clustering methods, *k*-means and self-organized map (SOM) [10]. The DEFE method makes use of innate characteristics in a gene expression dataset and a feature pattern directly carries the transcriptomics profile of the membership genes specific to the solution of a research problem. This method provides integral results across multiple pairwise comparisons and has been successfully applied to several research projects, including the case study presented herein.

The DEFE patterns should not be mistaken as merely an alternative to a Venn diagram. Actually, DEFE patterns provide more precise information beyond a Venn diagram in more complexed combination of multiple differential expression analyses. For example, a DEFE pattern contains all information of up- and down-regulations across the multiple differential expression analyses, whereas such information is hard to present in a Venn diagram without confusion. In addition, high-dimensional information (e.g., >6) is hard to be visualized by a Venn diagram, but it is straightforward when looking at DEFE patterns. A common objective of a Venn diagram is to identify overlapping genes from multiple differential expression analyses, while DEFE provides overlapping genes, as well as specific genes, unique to one or more differential expression analyses as demonstrated in the case study below. Additionally, a combination of multiple DEFE schemes can provide more precise discovery of specific groups of genes.

The complexity of DEFE grows exponentially as the length of a pattern string, B in 3^B, increases. This suggests that complexity both in time and in space increases when B grows higher. Nevertheless, the number of genes in the input dataset is finite; not all feature patterns will be of interest to a research problem and most patterns contain very few genes if they exist and can be ignored. In our earlier research, no genes are represented in nearly half of the 27 feature patterns when $B = 3$. In case of a large B, for example, $B = 8$ and 6 in [11], over 95% and 75% DEFE patterns have no genes. Thus, this approach allows us to focus on the top few highly frequent DEFE patterns. A certain number of specific feature patterns of interest, rather than the entire spectrum in the feature pattern space, would be provided at the early stage of a data analytics process. With regard to the time complexity, a modelling up to $B = 8$ in a feature pattern in [11] and the results were generated within a minutes for over 28,000 DEGs.

The DEFE method has been applied to extract informative differential expression feature patterns in earlier research on wheat Fusarium head blight (FHB) [11, 12], embryogenesis and grain development [13], and in the regulation of alternative splicing [14]. FHB is a devastating wheat disease caused by an ascomycetous fungus *Fusarium graminearum* and has frequent outbreaks in warm and humid or subhumid regions worldwide. In recent years, transcriptomic studies using microarray and RNA-seq have become effective strategies to identify differentially expressed genes as a result of the disease between FHB-resistant and FHB1-susceptible wheat genotypes and provide hints into molecular mechanisms of resistance [11, 12, 15]. Gene expression patterns are determined by a multitude of internal and external stimuli. The former is primarily related to tissue-specific developmental cues, while the latter can involve different abiotic and biotic stresses. Changes in gene expression patterns are often related to signaling pathways, including hormone signaling [11, 12, 15–18].

In our wheat FHB research, integration of multiple DEFE feature patterns enabled precision detection of genes potentially associated with FHB resistance generally and specifically to some FHB-resistant wheat genotypes and with FHB susceptibility [11] and genes enhanced or suppressed by a phytohormone during FHB infection [12]. The DEFE method enabled the discovery of transcriptomic activeness at the early stages during wheat embryogenesis [13]. The fundamental and usefulness of the DEFE method have been clearly demonstrated and described in Pan et al. [10]. We present in this chapter an R package together with a successful case study to illustrate its application in the transcriptomics research.

2 Materials

2.1 RNA-Seq Dataset

RNA-seq read counts of wheat (*Triticum aestivum*) from Pan et al. [11] are used in this study. The dataset consists of 48 samples collected at two time points, 2 and 4 days post inoculation (dpi) with *Fusarium graminearum*, from one FHB-susceptible (Shaw) and three FHB-resistant (HC374, Nyubai, and Wuhan 1) wheat genotypes, as described in [11]. The assembled raw read counts together with the RNA-seq data are available at the Gene Expression Omnibus of National Center for Biotechnology Information (NCBI-GEO) repository (GSE113128) (Table 1).

2.2 Computational Hardware

1. ×64-based processor with 32GB installed memory (RAM) (*see* **Note 1**).
2. Processor Intel® Core(TM) i7-4810MQ CPU at 2.80GHz, 4 cores, 8 logic processors (*see* **Note 1**).
3. Hard drive space 1 TB (*see* **Note 1**).

2.3 Computational Software

1. Microsoft Windows 10 Enterprise 64-bit operating system (*see* **Note 1**).
2. The complete case study analysis is performed within the R/Bioconductor environment. R (version ≥4.1.2) can be downloaded from the CRAN website (https://cran.r-project.org/) and installed on the local computer. Bioconductor (version ≥3.14) can be installed from the R shell:

Table 1
Forty-eight samples from four wheat genotypes in this case study

	Shaw	HC374	Nyubai	Wuhan 1
FHB	Susceptible	Resistant		
FHB 2 dpi	3	3	3	3
FHB 4 dpi	3	3	3	3
Water 2 dpi	3	3	3	3
Water 4 dpi	3	3	3	3

```
if (!require("BiocManager", quietly = TRUE))
    install.packages("BiocManager")
BiocManager::install(version = "3.14")
```

3. The R Studio Desktop version is available to download from https://www.rstudio.com/products/rstudio/download/ (*see* **Note 2**).
4. The DESeq2 R package [1] in Bioconductor can be installed from the R shell:

```
if (!requireNamespace("BiocManager", quietly =
   TRUE))
    install.packages("BiocManager")
BiocManager::install("DESeq2")
```

5. The DEFE R package developed by the authors (Supplementary File S1) can be downloaded and installed from the R shell (*see* **Note 3**), following **steps 6–8**.
6. Install supporting R packages:

```
install.packages(c("optparse", "plyr", "gtools",
      "tidyverse", "ggplot2", "reshape2"),
      dependencies=T)
```

7. Retrieve and install `DEFE.tar.gz` file from https://github.com/YPOtt/raw/main/DEFE:

```
install.packages("https://github.com/YPOtt/
DEFE/raw/main/DEFE.tar.gz", repos=NULL,
type="source")
```

8. RSEM to calculate transcript per million (TPM) values from BAM files [19]. TPM is an alternative to read counts to quantify extent of gene expression (*see* **Note 4**).

3 Methods

3.1 Obtaining the Raw Read Count Data

1. Download the raw read count data from GSE113128 (*see* **Note 5**).
2. Remove *Fusarium graminearum* genes from row 110,791 onward for the purpose of this case study focusing only on the wheat genes.
3. Save the file as tab delimited in the `caseStudy` folder as `rawReadCounts.txt`.

3.2 Perform Gene Differential Expression Analysis Using DESeg2

1. Load DESeq2 library (*see* **Note 6**):

```
library("DESeq2")
```

2. Set job working directory: insert path here to the working directory, for example, "E:/DEFE/caseStudy"

```
wd <- "E:/DEFE/caseStudy" #insert actual path
   here
setwd(wd)
```

3. Read input files (*see* **Notes 5** and **7**). Parameter files (`meta_-data.txt` and `DEAs.txt`) as shown in Tables 2 and 3 are available in Supplementary File S1.

```
# Read raw read counts:
ctsData <-
    as.data.frame(read.delim("rawReadCounts.
    txt", header=T, row.names=1, as.is=T,
    quote="", sep="\t"),stringsAsFactors=F)

# Read meta data indicating replicate of the
       same samples:
meta_data <- as.data.frame(read.table("me-
     ta_data.txt", header=T, row.names=1,
     as.is=T, quote="", sep="\t"), string-
     sAsFactors=F)

# Read file for differential expression ana-
       lysis (DEA) indicating the pairs of
       samples for comparison:
DEAs <- as.data.frame(read.table("DEAs.txt",
     header=T, row.names=1, as.is=T,
     quote="", sep="\t"), stringsAsFac-
     tors=F)
```

4. Separate the DEAs matrix into three vectors to fit DESeq2 function call:

```
Expts <- DEAs$expts
Ctrls <- DEAs$ctrls
exptIDs <- rownames(DEAs)
```

5. As a requirement of DESeq2, construct a data frame as follows:

```
meta_data <- data.frame(condition=meta_data
     $Condition, libType=meta_data$libType,
     row.names=colnames(ctsData))
```

Table 2
The content of "`meta_data.txt`" file for DESeq2 analysis

Sample	Condition	libType
HCFg_2d1	HCFg_2d	Mixed
HCFg_2d2	HCFg_2d	Mixed
HCFg_2d3	HCFg_2d	Mixed
HCFg_4d1	HCFg_4d	Mixed
HCFg_4d2	HCFg_4d	Mixed
HCFg_4d3	HCFg_4d	Mixed
HCH20_2d1	HCH20_2d	Mixed
HCH20_2d2	HCH20_2d	Mixed
HCH20_2d3	HCH20_2d	Mixed
HCH20_4d1	HCH20_4d	Mixed
HCH20_4d2	HCH20_4d	Mixed
HCH20_4d3	HCH20_4d	Mixed
NFg_2d1	NFg_2d	Mixed
NFg_2d2	NFg_2d	Mixed
NFg_2d3	NFg_2d	Mixed
NFg_4d1	NFg_4d	Mixed
NFg_4d2	NFg_4d	Mixed
NFg_4d3	NFg_4d	Mixed
NH20_2d1	NH20_2d	Mixed
NH20_2d2	NH20_2d	Mixed
NH20_2d3	NH20_2d	Mixed
NH20_4d1	NH20_4d	Mixed
NH20_4d2	NH20_4d	Mixed
NH20_4d3	NH20_4d	Mixed
SFg_2d1	SFg_2d	Mixed
SFg_2d2	SFg_2d	Mixed
SFg_2d3	SFg_2d	Mixed
SFg_4d1	SFg_4d	Mixed
SFg_4d2	SFg_4d	Mixed
SFg_4d3	SFg_4d	Mixed
SH20_2d1	SH20_2d	Mixed

(continued)

Table 2 (continued)

Sample	Condition	libType
SH20_2d2	SH20_2d	Mixed
SH20_2d3	SH20_2d	Mixed
SH20_4d1	SH20_4d	Mixed
SH20_4d2	SH20_4d	Mixed
SH20_4d3	SH20_4d	Mixed
WFg_2d1	WFg_2d	Mixed
WFg_2d2	WFg_2d	Mixed
WFg_2d3	WFg_2d	Mixed
WFg_4d1	WFg_4d	Mixed
WFg_4d2	WFg_4d	Mixed
WFg_4d3	WFg_4d	Mixed
WH20_2d1	WH20_2d	Mixed
WH20_2d2	WH20_2d	Mixed
WH20_2d3	WH20_2d	Mixed
WH20_4d1	WH20_4d	Mixed
WH20_4d2	WH20_4d	Mixed
WH20_4d3	WH20_4d	Mixed

6. Organize the count data into DESeq dataset format:

```
deseq_dataset <- DESeqDataSetFromMatrix
    (countData = ctsData, colData = me-
    ta_data, design = ~ condition)
```

7. Initialize three data matrices for holding result from each differential expression analysis:

```
log2FCs <- data.frame(row.names=rownames
  (ctsData))
pVals <- data.frame(row.names=rownames
  (ctsData))
pAdjs <- data.frame(row.names=rownames
  (ctsData))
```

Table 3
The content of "DEAs.txt" file for DESeq2 analysis

exptIDs	expts	ctrls
S_2d	SFg_2d	SH20_2d
S_4d	SFg_4d	SH20_4d
H_2d	HCFg_2d	HCH20_2d
H_4d	HCFg_4d	HCH20_4d
N_2d	NFg_2d	NH20_2d
N_4d	NFg_4d	NH20_4d
W_2d	WFg_2d	WH20_2d
W_4d	WFg_4d	WH20_4d
Fg_H_2d/S	HCFg_2d	SFg_2d
Fg_H_4d/S	HCFg_4d	SFg_4d
Fg_N_2d/S	NFg_2d	SFg_2d
Fg_N_4d/S	NFg_4d	SFg_4d
Fg_W_2d/S	WFg_2d	SFg_2d
Fg_W_4d/S	WFg_4d	SFg_4d
Wt_H_2d/S	HCH20_2d	SH20_2d
Wt_H_4d/S	HCH20_4d	SH20_4d
Wt_N_2d/S	NH20_2d	SH20_2d
Wt_N_4d/S	NH20_4d	SH20_4d
Wt_W_2d/S	WH20_2d	SH20_2d
Wt_W_4d/S	WH20_4d	SH20_4d

8. Execute the differential expression analysis:

```
    for (i in 1:length(DEAs$expts))
    {
        ctrl <- DEAs$ctrls[i]
        expt <- DEAs$expts[i]
        exptID <- exptIDs[i]
deseq_dataset$condition <-
      relevel(deseq_dataset$condition, ctrl,
          expt)
# call DESeq
deg <- DESeq(deseq_dataset)
deg_results <- results(deg, alpha=0.01,
    contrast=c("condition", expt, ctrl))
# add result of each iteration to result ma-
      trices
```

```
names(deg_results) <- paste(exptID,
   names(deg_results), sep=":")
   log2FCs <- data.frame(log2FCs, deg_results
      [2])
   pVals <- data.frame(pVals, deg_results[5])
   pAdjs <- data.frame(pAdjs, deg_results[6])
} # end of for
```

9. Prepare and save result. The result file contains raw and normalized read counts, normalized base means, and result of 20 differential expression analyses (log2 fold changes, p values and adjusted p values):

```
# get the raw and normalized read counts
count_table_original <- counts(deg, normali-
     zed=FALSE)
count_table_normalized <- round(counts(deg,
     normalized=TRUE), 4)

# summary of results containing raw and nor-
     malized read counts
count_table_deg_result <-
     data.frame(count_table_original,
     count_table_normalized, deg_results[1],
     log2FCs, pVals, pAdjs)

# save results
write.table(count_table_deg_result,
     file="results.txt", sep="\t", row.name-
     s=TRUE, col.name=NA, quote=FALSE)
```

3.3 Perform DEFE Analyses

Differential expression behavior of a gene at each pairwise comparison is modelled by three numerical values: 1 = up, 2 = down, and 0 = unchanged. The differential expression profile of a gene across a series of B conditions is modelled by concatenating the behaviors of the B pairwise comparisons in the series. For example, in a differential expression dataset consisting of B experimental conditions, B characters are used to describe the gene differential expression profile across the B conditions as they are compared between the treatments and their respective controls. The input data file to the DEFE algorithm consists of three blocks: (1) the expression values, usually presented as the means of normalized replicates for each experimental condition, (2) the log2 fold changes, and (3) the significance measures (either p or adjusted p value), which can be combined into a single input file (e.g., a revised result file from DESeq2 above: "results.txt"). The input file is saved as a tab

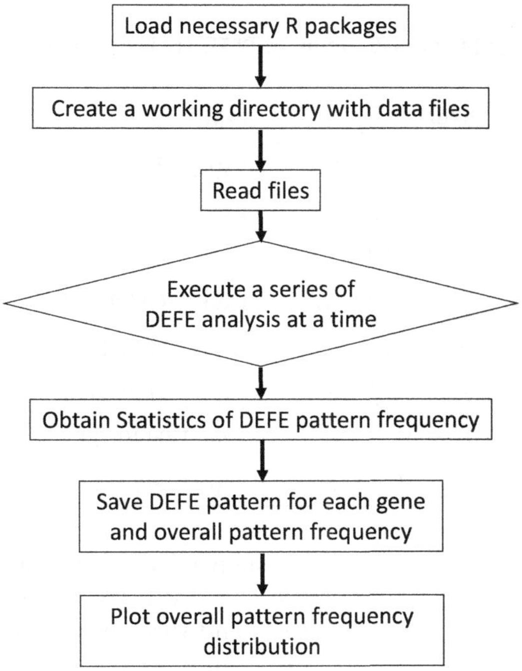

Fig. 1 Procedure diagram of DEFE operation

delimited file, such as the `DEGdata.txt` file in Supplementary File S2 used in this study. The second step is to identify the behavior of differential expression for each pairwise comparison based on the user's input. When the differential expression behaviors of each gene across *B* pairwise comparison are defined, a DEFE pattern is constructed by concatenating the behaviors of differential expression of the gene across the entire series of *B* pairwise comparisons and appended to the pattern vector (*V*). Finally, the number of genes affiliated with each DEFE patterns is generated and features (genes) with the interested DEFE pattern are extracted from the dataset. An R package is available at https://github.com/YPOtt/DEFE. An example data file and three parameter files for this case study are provided in Supplementary File S2. The operation procedures are shown in Fig. 1. The steps for the case study are presented below.

1. Load the DEFE library together with its supporting libraries:

```
library("optparse")
library("plyr")
library("gtools")
library("tidyverse")
library("ggplot2")
library("reshape2")
library("DEFE")
```

Table 4
The parameter for the FW series of DEFE analysis, same as the "defeSeriesFW.txt" file

exptID	expr1	expr2	log2FC	padj
FW2_S	Expr_SFg_2d	Expr_SH20_2d	Log2FC_FW2_S	pAdj_FW2_S
FW4_S	Expr_SFg_4d	Expr_SH20_4d	Log2FC_FW4_S	pAdj_FW4_S
FW2_HC	Expr_HCFg_2d	Expr_HCH20_2d	Log2FC_FW2_HC	pAdj_FW2_HC
FW4_HC	Expr_HCFg_4d	Expr_HCH20_4d	Log2FC_FW4_HC	pAdj_FW4_HC
FW2_Nb	Expr_NFg_2d	Expr_NH20_2d	Log2FC_FW2_Nb	pAdj_FW2_Nb
FW4_Nb	Expr_NFg_4d	Expr_NH20_4d	Log2FC_FW4_Nb	pAdj_FW4_Nb
FW2_Wh	Expr_WFg_2d	Expr_WH20_2d	Log2FC_FW2_Wh	pAdj_FW2_Wh
FW4_Wh	Expr_WFg_4d	Expr_WH20_4d	Log2FC_FW4_Wh	pAdj_FW4_Wh

2. Put all needed files in a job working directory (e.g., `E:/DEFE/caseStudy`). These include the DEG data file (`DEGdata.txt`) and three parameter files (`defeSeriesFW.txt`, `defeSeriesFRS.txt`, and `defeSeriesWRS.txt`) available in Supplementary File S2. Set job working directory:

```
wd <- "E:/DEFE/caseStudy" #insert actual path
    here
setwd(wd)
```

3. Read input files (*see* **Note 8**):

```
input.filename <- "DEGdata.txt"
data.m <- as.data.frame(read.delim(input.fi-
     lename, header=T, row.names=1, as.is=T,
     quote="", sep="\t"), stringsAsFac-
     tors=F)
```

4. Read parameter files. Three series of DEFE analyses are performed to model and are itemized in **steps 5–7** (*see* **Note 8**).

5. The differential expression of each treatment as compared to concurrent water treatment samples: series FW, where "F" denotes the treatments of *Fusarium graminearum* to a given genotype and sample taken at 2 or 4 dpi and "W" the corresponding water treatments (Table 4):

```
FW.params.m <- get.user.parameters("defe-
   SeriesFW.txt")
```

6. Differential expression in response to FHB between the resistant genotypes (HC374, Nyubai, and Wuhan 1) and the susceptible Shaw: series FRS, where "R" denotes a resistant genotype and "S" susceptible Shaw (Table 5):

Table 5
The parameter for the FRS series of DEFE analysis, same as the "`defeSeriesFRS.txt`" file

exptID	expr1	expr2	log2FC	padj
FRS2_HC	Expr_HCFg_2d	Expr_SFg_2d	Log2FC_FRS2_HC	pAdj_FRS2_HC
FRS4_HC	Expr_HCFg_4d	Expr_SFg_4d	Log2FC_FRS4_HC	pAdj_FRS4_HC
FRS2_Nb	Expr_NFg_2d	Expr_SFg_2d	Log2FC_FRS2_Nb	pAdj_FRS2_Nb
FRS4_Nb	Expr_NFg_4d	Expr_SFg_4d	Log2FC_FRS4_Nb	pAdj_FRS4_Nb
FRS2_Wh	Expr_WFg_2d	Expr_SFg_2d	Log2FC_FRS2_Wh	pAdj_FRS2_Wh
FRS4_Wh	Expr_WFg_4d	Expr_SFg_4d	Log2FC_FRS4_Wh	pAdj_FRS4_Wh

Table 6
The parameter for the WRS series of DEFE analysis, same as "`defeSeriesWRS.txt`" file

exptID	expr1	expr2	log2FC	padj
WRS2_HC	Expr_HCH20_2d	Expr_SH20_2d	Log2FC_WRS2_HC	pAdj_WRS2_HC
WRS4_HC	Expr_HCH20_4d	Expr_SH20_4d	Log2FC_WRS4_HC	pAdj_WRS4_HC
WRS2_Nb	Expr_NH20_2d	Expr_SH20_2d	Log2FC_WRS2_Nb	pAdj_WRS2_Nb
WRS4_Nb	Expr_NH20_4d	Expr_SH20_4d	Log2FC_WRS4_Nb	pAdj_WRS4_Nb
WRS2_Wh	Expr_WH20_2d	Expr_SH20_2d	Log2FC_WRS2_Wh	pAdj_WRS2_Wh
WRS4_Wh	Expr_WH20_4d	Expr_SH20_4d	Log2FC_WRS4_Wh	pAdj_WRS4_Wh

```
FRS.params.m <-
    get.user.parameters("defeSeriesFRS.
    txt")
```

7. Background differential in response to water treatment between the resistant genotypes (HC374, Nyubai, and Wuhan 1) and the susceptible Shaw: series WRS (Table 6):

```
WRS.params.m <-
    get.user.parameters("defeSeriesWRS.
    txt")
```

8. Perform DEFE analyses of data from **step 3** using defined parameter setting in the data from **steps 5**, **6**, and **7**: the thresholds are abs(log2FC) ≥ 2, adjusted $p \leq 0.01$, and max (Expr1, Expr2) ≥ 100 (*see* **Note 8**):

```
#FW series of DEFE analysis
FW.data.m.pattern <-
  get.DEFE("FW",FW.params.m,data.
      m,2,0.01,100)
```

```
FW.pattern.Vec <- get.pattern.vec(FW.data.m.
      pattern)

#FRS series of DEFE analysis
FRS.data.m.pattern <- get.DEFE("FRS",FRS.
      params.m,data.m,2,0.01,100)
FRS.pattern.Vec <- get.pattern.vec(FRS.
      data.m.pattern)
#WRS series of DEFE analysis
WRS.data.m.pattern <- get.DEFE("WRS",WRS.
      params.m,data.m,2,0.01,100)
WRS.pattern.Vec <- get.pattern.vec(WRS.
      data.m.pattern)
```

9. Obtain the statistics of the DEFE pattern frequency:

```
FW.pattern.freq <- get.pattern.freq(FW.pat-
      tern.Vec)
FRS.pattern.freq <- get.pattern.freq(FRS.
      pattern.Vec)
WRS.pattern.freq <- get.pattern.freq(WRS.
      pattern.Vec)
```

10. Save the DEFE pattern result for each gene (Supplementary File S3) (*see* **Note 9**):

```
To_be_saved <- data.frame(data.m, FW.data.m.
      pattern, FRS.data.m.pattern, WRS.data.-
      m.pattern, FW.pattern.Vec, FRS.pattern.
      Vec, WRS.pattern.Vec)
write.table(To_be_saved, file="DEFEresults.
      txt", sep="\t", row.names=TRUE, col.
      name=NA, quote=FALSE)
```

11. Save the statistics of the DEFE pattern frequency of each series of analysis (Supplementary File S3):

```
#save frequencies of FW series patterns
write.table(FW.pattern.freq, file="FW_DEFE_-
      PatternFreq.txt", sep="\t", row.name-
      s=TRUE, col.name=NA, quote=FALSE)

#save frequencies of FRS series patterns
write.table(FRS.pattern.freq, file="FRS_DE-
      FE_PatternFreq.txt", sep="\t", row.
      names=TRUE, col.name=NA, quote=FALSE)
```

Table 7
Three series of DEFE analyses and the number of DEGs associated with each series

Series prefix	Modelled for	Treatment	DEFE series[a]	DEGs
FW	Effect of FHB challenge	Fg[b] vs. water[c]	FW(S_2d, S_4d, H_2d, H_4d, N_2d, N_4d, W_2d, W_4d)	10,957
FRS	FHB resistant versus susceptible plants	Fg	FRS(H/S_2d, H/S_4d, N/S_2d, N/S_4d, W/S_2d, W/S_4d)	8792
WRS		Water	WRS(H/S_2d, H/S_4d, N/S_2d, N/S_4d, W/S_2d, W/S_4d)	2221

[a] Explanation of each series of comparisons: 1. Comparison between each *F. graminearum*-treated sample and the corresponding water-treated control sample, where the prefix "FW" stands for pairwise comparison between *F. graminearum*-treated samples (F) and water-treated samples (W), the numerical character "1" denotes "up-" and "2" "down-" regulation by *F. graminearum* treatment; for each comparison, S stands for Shaw, H for HC374, N for Nyubai, and W for Wuhan 1 2. Comparisons of FHB-resistant HC374, Nyubai, and Wuhan 1 with the susceptible Shaw in *F. graminearum*-treated plants (FRS) and in water mock-treated plants (WRS), respectively, where the second letter "R" in the prefix stands for FHB resistance and the third letter "S" for FHB susceptible

[b] Stands for *F. graminearum*-treated samples

[c] Stands for water-treated samples

```
#save frequencies of WRS series patterns
write.table(WRS.pattern.freq, file="WRS_DE-
      FE_PatternFreq.txt", sep="\t", row.
      names=TRUE, col.name=NA, quote=FALSE)
```

12. Collectively, 13,373 DEGs were identified in one or more series of the DEFE analyses (Table 7).

13. Plot and save the image of the top 50 DEFE pattern frequency distributions from the three series of analyses (Fig. 2) (*see* **Notes 10** and **11**):

```
#from FW series
freq.save.tiff(data.m.pattern= FW.data.m.
      pattern, out.filename="FW_patternfeq-
      diagram.tiff", rmsubset=50, color="-
      black")

#from FRS series
freq.save.tiff(data.m.pattern=FRS.data.m.
      pattern, out.filename="FRS_patternfeq-
      diagram.tiff", rmsubset=50, color="-
      black")
```

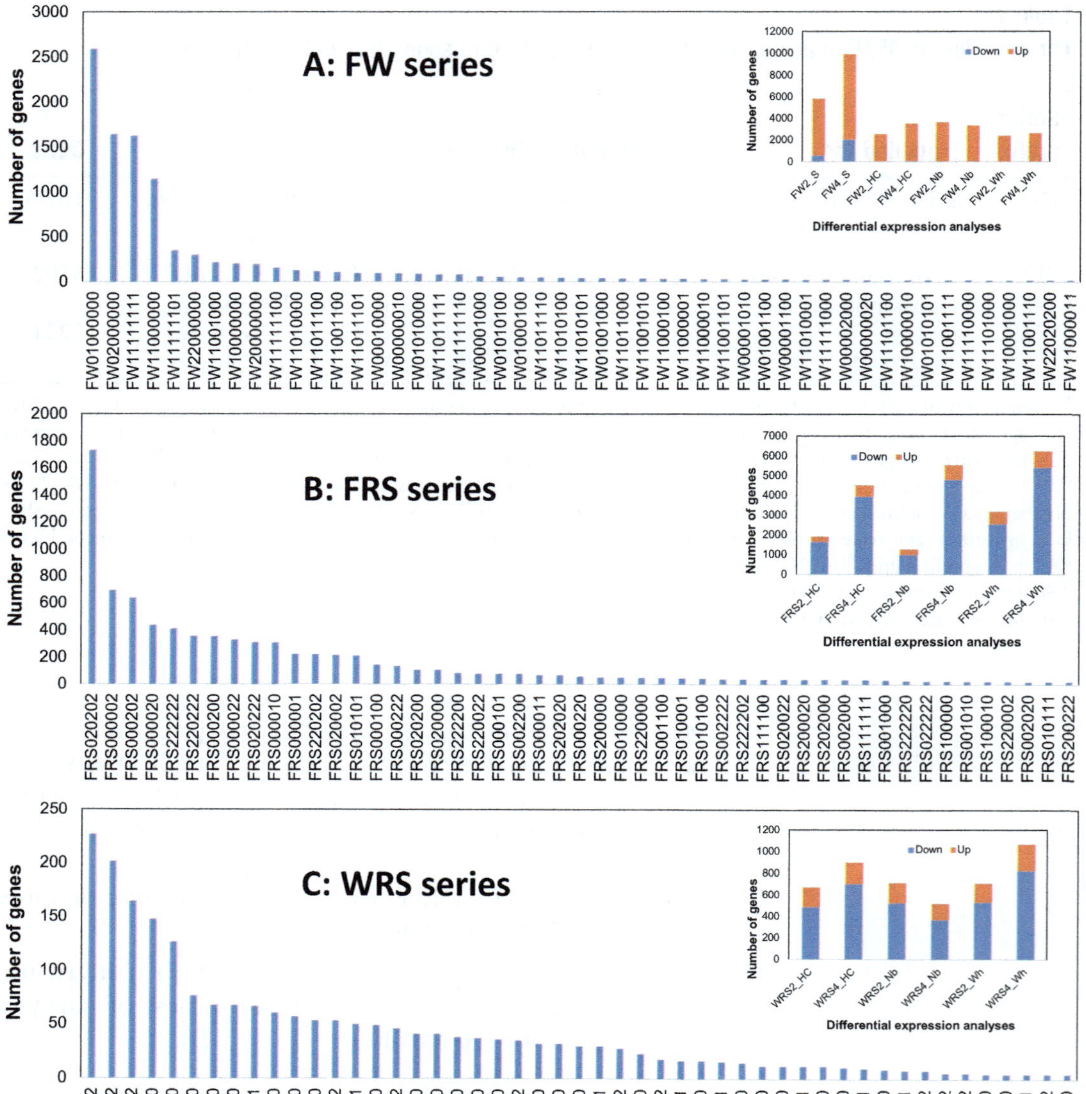

Fig. 2 Frequency distribution of the top 50 DEFE patterns of FW (**a**), FRS (**b**), and WRS (**c**) series. The insert in each panel indicates the number of up- and downregulated genes of each pairwise comparison in the DEFE pattern in the same order. In the insert panel, *UP* upregulation, *DN* downregulation, *F Fusarium* treated, *W* water treated, *FW Fusarium* treated over water treated, *R* FHB-resistant genotypes, *S* FHB-susceptible genotype, *_S* superb, *_HC* HC174, *_Nb* Nyubai, *Wh* Wuhan 1, *FRS* a resistant genotype over the susceptible Shaw in *Fusarium*-treated samples, *WRS* a resistant genotype over the susceptible Shaw in water-treated samples; the digital character after "FRS" or "WRS" denotes the samples were taken at 2 or 4 dpi

```
#from WRS series
 freq.save.tiff(data.m.pattern=WRS.data.m.
      pattern, out.filename="WRS_pattern-
      feqdiagram.tiff", rmsubset=50, col-
      or="black")
```

3.4 Perform Functional Group Analyses

3.4.1 To Identify Groups of Genes Associated with FHB Resistance

1. The logic is that this group of genes would have to be upregulated by FHB in all three FHB-resistant genotypes at both time points, but not or downregulated in FHB-susceptible Shaw. In this regard, we try to find the group of genes having a DEFE pattern of (*see* **Note 12**):

   ```
   FW00111111 has 2 genes.
   FW20111111 has 0 gene.
   FW02111111 has 0 gene.
   ```

2. These genes would have to expressed significantly higher in the three resistant genotypes than in the susceptible Shaw at both time points:

   ```
   FRS111111 has 36 genes.
   ```

3. But these group of genes should not be up regulated in the control sample of any genotype at any time point, such patterns:

   `WRS######` (where # denotes a character of either 1 or 0, but not all six "#" are 0) have 509 genes collectively.

4. With `FW00111111∩(FRS111111¬WRS######)` combined DEFE feature patterns, we confirm that the two genes having `FW00111111` also have `FRS111111`, but not `WRS######`. We consider these two genes to be putatively associated with FHB resistance (*see* **Note 13**).

3.4.2 To Identify Groups of Genes Associated with FHB Susceptibility

This section is similar to Subheading 3.4.1, but is performed in the opposite direction.

1. The logic is that these genes would have to be up-regulated by FHB in the FHB susceptible genotype Shaw, but not or down-regulated in any of the three FHB resistant genotypes. In this regard, we try to find the group of genes having DEFE pattern of:

   ```
   FW11000000 has 1151 genes.
   FW11****** (where each * denotes either 0 or 2) have
   no gene.
   ```

2. These genes would have to expressed significantly higher in the susceptible Shaw than in all three resistant genotypes at both time points:

   `FRS222222 has 414 genes.`

3. But these group of genes should not be in the control at any of the six pairwise comparisons in water sample where Shaw expressed higher than any of the three FHB resistant genotypes (i.e. no "2" appears in the WRS pattern):

   `WRS###### (where # denotes either 0 or 2, but not all six are 0) have 1682 genes collectively.`

4. Combining `FW11000000∩(FRS222222¬WRS######)` DEFE feature patterns, we obtain **193** genes (*see* **Note 14**).

3.5 Technical Insights

This study demonstrated that DEFE is a practical and unique pattern recognition method to extract gene expression patterns closely relevant to a research objective. DEFE uses the innate features in the data and surpasses what Venn diagram can deliver in pattern recognition by (1) revealing combinations of up- and downregulation in the same DEFE pattern, (2) identifying high numbers of pairwise differential expression analyses, and (3) enabling integration of multiple schemes of DEFE analyses. With these advantages, the DEFE method can facilitate transcriptomic analyses in different crops or organisms and has already been successfully applied in numerous studies in wheat, which is an allohexaploid crop species with complex genetics and a large (17-Gb) genome size [25], including analyses of the FHB disease response [11, 12], embryogenesis and grain development [13], the regulation of alternative splicing [14], and discovery of genes closely associated with cold acclimation and otherwise [26].

4 Notes

1. These are suggested parameters and guidelines; however, any computer operating system with R installed would be suitable. The R package provided for this case study has been tested in both MS Window and Linux.
2. R Studio is convenient for operation, such as code modification, but not essential for the job. Directly using R (> = 4.1.2) will be fine.
3. It is important to install all supporting R packages and load into R session to perform DEFE analyses; otherwise, DEFE would not perform properly. Installation of the DEFE package can be

done from the working directory "wd" if it is specified. For example,

```
wd <- D:/DEFE
setwd(wd)
install.packages("DEFE.tar.gz", repos = NULL,
     type = "source")
```

4. The expression data are in read counts or transcripts per million (TPM); for this case study, read counts is used. A tool such as [21] can be used to calculate the TPM value from the BAM file generated by STAR [20].
5. The input data for DESeq2 should contain a matrix of integer values of unnormalized read counts or estimated counts of sequencing reads (for single-end RNA-seq) or fragments (for paired-end RNA-seq). Normalized expression data, such as TPM, FPKM, or RPKM, is not acceptable for DEG analysis [4, 5]. The data is saved as plain text file, formatted as either the tab delimited (*.txt or *.tab) or comma separated value (*.csv), and read by corresponding R readers.
6. The updated DESeq2 operation instruction is available at http://bioconductor.org/packages/release/bioc/vignettes/DESeq2/inst/doc/DESeq2.html
7. User may choose to enter meta data and parameters for differential expression analysis manually. It is recommended to enter from file so that it is well prepared and has a track record of analyses. For this case study, the meta data file is in Table 2 and the DEAs file is in Table 3.
8. For definition of differentially expressed genes (DEGs), the absolution value of log2 fold change (e.g., ≥ 1 or ≥ 2) and a statistical significance measure, p-values or adjusted p-values (e.g., ≤ 0.05 or ≤ 0.01), are usually applied. The use of p-values or adjusted p-values together with a log2 fold change is usually sufficient to justify regular DEG analysis. However, in order to mitigate the impact of noise, especially in a highly noisy dataset, such as in this case study of wheat infected with *F. graminearum*, an additional criterion, namely, significant expression level, can also be applied. That is to say, within the pair of samples being compared for a given gene, the expression level of at least one sample has to be significant. There is no universal definition of this significance level. Based on our experience and from literature, a read count of ≥ 50 (we used ≥ 100 read counts in the case study to deal with the data noise) or TPM ≥ 2 is considered significant.

In case of performing DEFE analysis without using expression values, for example, in mammalian RNA-seq analysis, a simpler function "`get.DEFE.no.expr()`" should be called:

```
data.m.pattern <- get.DEFE.no.expr(prefix,
      params.m, data.m, log2FG.threshold,
      pVal.threshold)
```

For example, we use the FW series of DEFE analysis.

```
F2W.params.m = FW.params.m[,3:4]
F2W.data.m.pattern = get.DEFE.no.expr("F2W",
     F2W.params.m, data.m, 2, 0.01)
F2W.pattern.Vec = get.pattern.vec(F2W.data.-
     m.pattern)
F2W.pattern.freq = get.pattern.freq(F2W.pat-
     tern.Vec)
To_be_saved <- data.frame(F2W.data.m.pat-
     tern, F2W.pattern.Vec)
write.table(To_be_saved, file="F2Wresults.
     txt", sep="\t", row.names=TRUE, col.
     name=NA, quote=FALSE)

write.table(F2W.pattern.freq, file="F2W_DE-
      FE_PatternFreq.txt", sep="\t", row.
      names=TRUE, col.name=NA, quote=FALSE)

freq.save.png(F2W.data.m.pattern, "F2W_pat-
     ternfeqdiagram_png", rmsubset=60, col-
     or="blue")
```

9. The results to be saved depend on user's wish on what to be saved. For example, since the input data is already available and the user may not want to keep it in the result file, the `To_be_-saved` matrix can be organized as follows:

```
To_be_saved <- data.frame(FW.data.m.pattern,
     FW.pattern.Vec)
```

If the user prefers to save only the DEFE pattern associated with each gene without the details of up- or downregulation as each pairwise comparison:

```
To_be_saved <- data.frame(FW.pattern.Vec)
```

10. While saving the image, there might be a warning message in red, such as:

```
No id variables; using all as measure vari-
   ables
null device
          1
```

There is no concern to such warning.

11. The DEFE R package offers all four most popular formats of image files to be saved: BMP, JPEG, PNG, and TIFF. Users are free to specify at the postscript of the function name as well as the image file extension. Users are also free to choose the top number of patterns and color of the bars and wish to save a specific type of file. When a series of DEFE analyses generate a less number of DEFE patterns, the software will plot whichever is less. A barplot of more than 70 top patterns is not recommended as the bars will be crowded with no separation at that high number, for example, when a user wants to plot frequencies of top 30 patterns, bar color in "blue" and saved as a "`*.png`" file:

```
freq.save.png(data.m.pattern=FW.data.m.pat-
     tern, out.filename="FW_patternfeqdia-
     gram_png", rmsubset=30, color="blue")
```

12. The frequency number of each DEFE pattern in this case is different from Pan et al. [11] because a higher stringency of significant differential expression (abs(log2FC) $\geq$ 2) was employed here.
13. The two genes (TraesCS7A01G021900, TraesCS7D01G019400) found in this case study to be putatively associated with FHB resistance are both glutathione S-transferase (GST). Although upregulation of GSTs has been observed in FHB-resistant as well as FHB-susceptible wheat material following FHB infection, increased levels of specific GST family members have been observed in wheat and barley in association with FHB resistance [9, 20–22]. GSTs are known for their roles in maintaining the physiological redox state of the cell, protection of the cell against oxidative damages, and detoxification of xenobiotics [23, 24].
14. Three genes in this group of 193 genes putatively associated with FHB susceptibility were not considered in this group in Pan et al. [11] because a lesser extent of upregulation by FHB was observed among the three FHB-resistant genotypes in one or both time points. These 193 genes are available in Supplementary File S4. A detailed discussion is available in [11].

Acknowledgments

This work was supported by the Canadian Wheat Improvement program of the National Research Council Canada, Genomic Research and Development Initiative of Agriculture Agri-Food Canada, and Canadian Wheat Alliance. We thank Yifeng Li, Lipu Wang, Ravinder Goyal, and Pierre Fobert for their involvement in the discussion at the early stage of the development of the DEFE method.

References

1. Love MI, Huber W, Anders S (2014) Moderated estimation of fold change and dispersion for RNA-seq data with DESeq2. Genome Biol 15:550
2. Trapnell C, Hendrickson DG, Sauvageau M et al (2013) Differential analysis of gene regulation at transcript resolution with RNA-seq. Nature Biotechnol 31:46–53
3. Robinson MD, McCarthy DJ, Smyth GK (2010) edgeR: a Bioconductor package for differential expression analysis of digital gene expression data. Bioinform 26:139–140
4. Zhao S, Ye Z, Stanton R (2020) Misuse of RPKM or TPM normalization when comparing across samples and sequencing protocols. RNA 26:903–909
5. Zhao Y, Li MC, Konaté MM et al (2021) TPM, FPKM, or normalized counts? A comparative study of quantification measures for the analysis of RNA-seq data from the NCI patient-derived models repository. J Transl Med 19:269
6. Rousseeuw PJ (1987) Silhouettes: a graphical aid to the interpretation and validation of cluster analysis. J Comp Appl Math 20:53–65
7. Famili AF, Liu G, Liu Z (2004) Evaluation of optimization of clustering in gene expression data analysis. Bioinformatics 20:1535–1545
8. Pan Y, Pylatuik JD, Ouyang J et al (2004) Discovery of functional genes for systemic acquired resistance in *Arabidopsis thaliana* through integrated data mining. J Bioinforma Comput Biol 2:639–655
9. Tchagang AB, Pan Y, Famili F et al (2013) Biclustering of DNA microarray data: theory, evaluation, and applications. In: Palshikar GK (ed) Bioinformatics: concepts, methodologies, tools, and applications, chapter 29. IGI Global
10. Pan Y, Li Y, Liu Z et al (2019) Differential expression feature extraction (DEFE) and its application in RNA-seq data analysis. bioRxiv. https://doi.org/10.1101/511188
11. Pan Y, Liu Z, Rocheleau H et al (2018) Transcriptome dynamics associated with resistance and susceptibility against fusarium head blight in four wheat genotypes. BMC Genomics 19:642
12. Buhrow LM, Liu Z, Cram D et al (2021) Wheat transcriptome profiling reveals abscisic and gibberellic acid treatments regulate early-stage phytohormone defense signaling, cell wall fortification, and metabolic switches following *Fusarium graminearum*-challenge. BMC Genomics 22:798
13. Xiang D, Quilichini T, Liu Z et al (2019) The transcriptional landscape of polyploid wheats and their diploid ancestors during embryogenesis and grain development. Plant Cell 31:2888–2911
14. Gao P, Quilichini TD, Zhai C et al (2021) Alternative splicing dynamics and evolutionary divergence during embryogenesis in wheat species. Plant Biotechnol J 19:1624–1643
15. Brauer EK, Rocheleau H, Balcerzak M et al (2019) Transcriptional and hormonal profiling of *Fusarium graminearum*-infected wheat reveals an association between auxin and susceptibility. Physiol Mol Plant Pathol 107:33–39
16. Bari R, Jones JDG (2009) Role of plant hormones in plant defence responses. Plant Mol Biol 69:473–488
17. Peleg Z, Blumwald E (2011) Hormone balance and abiotic stress tolerance in crop plants. Curr Opin Plant Biol 14:290–295
18. Wang L, Li Q, Liu Z et al (2018) Integrated transcriptome and hormone profiling highlight the role of multiple phytohormone pathways in wheat resistance against fusarium head blight. PLoS One 13:e0207036
19. Li B, Dewey CN (2011) RSEM: accurate transcript quantification from RNA-Seq data with or without a reference genome. BMC Bioinform 12:323

20. Dobin A, Davis CA, Schlesinger F et al (2013) STAR: ultrafast universal RNA-seq aligner. Bioinform 29:15–21
21. RSEM-calculate-expression – estimate gene and isoform expression from RNA-Seq data. https://deweylab.github.io/RSEM/rsem-calculate-expression.html. Accessed on 28 Jan 2022
22. Foroud NA, Ouellet T, Laroche A et al (2012) Differential transcriptome analyses of three wheat genotypes reveal different host response pathways associated with Fusarium head blight and trichothecene resistance. Plant Pathol 61: 296–314
23. Gullner G, Kômives T (2006) Defense reactions of infected plants: roles of glutathione and glutathione S-transferase enzymes. Acta Phytopathol Entomol Hung 41:3–10
24. Dubreuil-Maurizi C, Poinssot B (2012) Role of glutathione in plant signaling under biotic stress. Plant Signal Behav 7:210–212
25. Brenchley R, Spannagl M, Pfeifer M et al (2012) Analysis of the bread wheat genome using whole-genome shotgun sequencing. Nature 491:705–710
26. Pan Y, Li Y, Liu Z, Zou J, Li Q (2022). Computational genomics insights into cold acclimation in wheat. Front Genet. 13: 1015673. https://doi.org/10.3389/fgene.2022.1015673

Chapter 12

Proteomic Profiling of Host Response in the Cereal Crop *Triticum aestivum* to the Mycotoxin, 15-Acetyldeoxynivalenol, Produced by the Fungal Pathogen, *Fusarium graminearum*

Reid Buchanan, Mitra Serajazari, and Jennifer Geddes-McAlister

Abstract

Deoxynivalenol (DON) is a destructive mycotoxin produced by the fungal pathogen *Fusarium graminearum* in the devastating cereal disease Fusarium head blight (FHB). Host resistance to FHB has been identified within some of these crops (e.g., wheat, barley, corn); however, identification of how the host reduces the production of, and tolerates, DON to lessen the effects of the disease still requires further discovery. The field of quantitative proteomics is an effective tool for measuring and quantifying host defense responses to external factors, including the presence of pathogens and toxins. Success within this area of research has increased through recent technological developments (e.g., instrument sensitivity) and the accessibility of data analysis programs. One advancement we leverage is the ability to label peptides with isobaric mass tags to allow for sample multiplexing, reducing mass spectrometer run times, and providing accurate quantification. In this protocol, we exemplify this methodology to identify protein-level responses to DON within both FHB-resistant and FHB-susceptible *Triticum aestivum* cultivars using tandem mass tags for quantitative labeling combined with liquid-chromatography-MS/MS (LC-MS/MS) analysis. Furthermore, this protocol can be extrapolated for the identification of host responses under various conditions, including infection and environmental fluctuations, to elucidate changes in proteomic profiling in diverse biological contexts.

Key words Mass spectrometry-based proteomics, *Triticum aestivum*, Deoxynivalenol, *Fusarium graminearum*, Tandem mass tags (TMT)-labeling, Mycotoxin, Bioinformatics

1 Introduction

Cereal crops, such as *Triticum aestivum*, *Hordeum vulgare*, and *Zea mays*, are important agricultural crops globally and, with the use of selective breeding strategies, grow in many environments. Despite adapted growth, these crops are sensitive to infection by fungal pathogens, requiring additional prevention and treatment methods to counteract destruction by disease [1]. Invading fungal

Nora A. Foroud and Jonathan A. D. Neilson (eds.), *Plant-Pathogen Interactions*, Methods in Molecular Biology, vol. 2659, https://doi.org/10.1007/978-1-0716-3159-1_12,

pathogens typically produce secondary metabolites (e.g., mycotoxins), which aid in the fight against host defense responses to infection and allow for enhanced virulence [1–3]. Deoxynivalenol (DON) is an example of a mycotoxin produced by the fungal pathogen, *Fusarium graminearum*, in the disease Fusarium head blight (FHB) [1, 2]. DON acts as a virulence factor for *F. graminearum* by binding to the ribosome and inhibiting protein synthesis within the host system [4]. This mode of action depletes susceptible hosts from creating new enzymes required to elicit an immune response to the pathogen, allowing the fungus to accumulate resources for growth and further infection. In addition to being destructive to the plant's health, DON is also toxic to mammals that consume infected plant material, causing vomiting, diarrhea, and weight loss [4, 5]. Previous research identifying resistance responses within *T. aestivum* to FHB define genes associated with facilitating host defense response to infection, such as *fhb1*, and although there have been studies into the identification of DON-resistance genes (e.g., those encoding ABC transporters, glutathione-S-transferases, and TaFROG), there is still a requirement to further elucidate the strategies in which the mycotoxin is managed [6–8].

Using a bottom-up proteomics approach, which relies on the enzymatic digestion of proteins into peptides from complex biological samples, the discovery of host proteins involved in resistance to FHB and protection from the accumulation of DON within *T. aestivum* have been explored [6, 9–13]. While label-free quantification strategies are applied to diverse sample sets, the benefits of combining protein extraction with chemical labeling, such as tandem mass tags (TMTs), permit multiplexing of samples for greater quantification accuracy and less instrumentation time [14, 15]. In this protocol, we outline inoculation of FHB-resistant and FHB-susceptible wheat varieties with purified DON, followed by an overview of our optimized protein extraction protocol combined with TMT labeling to identify and quantify changes in the plant host defense response in the presence of the mycotoxin with accuracy and reproducibility (Fig. 1). Overall, this procedure identifies proteins in the FHB-resistant, DON-inoculated *T. aestivum* samples involved in the plant's increased survivability to the mycotoxin. These findings are applicable for future breeding strategies to increase natural resistance to the devastating effects of FHB and possibly other mycotoxins.

2 Materials

All solutions listed below are synthesized using analytical grade reagents in UltraPure water (18MΩ at 25 °C) and are prepared fresh unless stated otherwise.

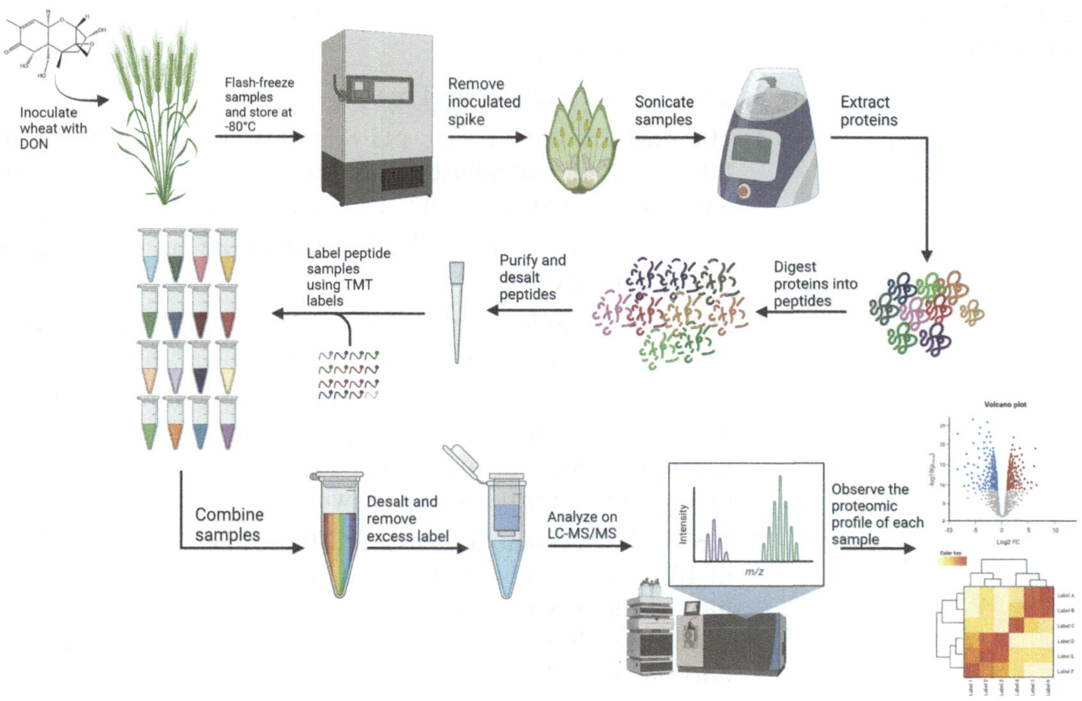

Fig. 1 Example workflow of bottom-up proteomics protocol using TMT labelling for DON inoculation in *T. aestivum*. Image created in BioRender.com

2.1 Mock and Deoxynivalenol Inoculation Solutions

1. 0.2% Tween-20 (*v*/*v*) mock inoculant.
2. 1.0 mg/mL 15-acetyldeoxynivalenol dissolved in 0.2% Tween-20.
3. 0.1 mg/mL 15-acetyldeoxynivalenol dissolved in 0.2% Tween-20.

2.2 Plant Material

1. FHB-resistant and FHB-susceptible *Triticum aestivum* seeds (*see* **Note 1**).
2. 1-gallon pots.
3. 1:1 mixture of nutrient-enriched planting mix and coarse planting clay.
4. Greenhouse or growth room capable of controlling temperatures between 18 °C and 21 °C and 16 h/8 h light cycles.
5. Greenhouse or growth room capable of controlling temperatures between 22 °C and 27 °C, humidity between 75% and 80%, and 16 h/8 h light cycles.
6. Glycerin crossing bag to cover wheat heads.

2.3 Proteome Extraction

1. 2 mL low protein binding microcentrifuge tubes.
2. Tissue homogenizer.
3. 0.9–2.0 mm stainless steel beads.
4. 100 mM Tris–HCl buffer, pH 8.5: Dissolve 10.9 g of Tris in 800 mL water and adjust to a pH of 8.5 using HCl, and then add water to a final volume of 1 L (stored at 4 °C).
5. Proteinase inhibition cocktail tablet (store as per supplier recommendations).
6. 20% sodium dodecyl sulfate (SDS): Dissolve 200 g of SDS in 1 L of water (store at room temperature).
7. 1 M dithiothreitol (DTT): In a fume hood, dissolve 0.781 g of DTT in 5 mL of water, and then aliquot into 1.5 mL sterile microfuge tubes (flash-frozen and stored at −20 °C). Discard of any unused portions after thawing.
8. 0.55 M iodoacetamide (IAA): In a fume hood, dissolve 2.03 g of IAA into an aluminum-foiled container, and then aliquot into sterile 1.5 mL microcentrifuge tubes (flash-frozen and stored at −20 °C). Keep out of light and discard of any unused portions after thawing.
9. 100% and 80% acetone (stored at −20 °C).
10. 8 M urea/40 mM HEPES: Dissolve 9.6 g urea and 0.19 g HEPES in 20 mL of water, and aliquot into sterile 1.5 mL microcentrifuge tubes (flash-frozen and stored at −20 °C).
11. 50 mM ammonium bicarbonate: Dissolve 0.60 g of ammonium bicarbonate in 150 mL of water (stored at room temperature).
12. Trypsin/Lys-C enzyme mixture (reconstituted as per manufacturer instructions).

2.4 Stop-and-Go Extraction (STAGE) Tip Desalting

1. Anhydrous acetonitrile (store at room temperature).
2. Stopping solution: 20% acetonitrile, 6% trifluoroacetic acid (store at room temperature).
3. Buffer A: 2% acetonitrile, 0.5% acetic acid, 0.1% trifluoroacetic acid (store at room temperature).
4. Buffer B: 80% acetonitrile, 0.5% acetic acid (store at room temperature).
5. Vacuum centrifuge.

2.5 TMT Labeling

1. Materials provided in the selected 16-plex TMT label reagent set (store desiccated at 4 °C) (*see* **Note 2**).
2. Anhydrous acetonitrile (store at room temperature).
3. 50% hydroxylamine (store at room temperature).

4. 100 mM HEPES pH 8.5: Add 2.38 g of HEPES to 80 mL of water and dissolve. Adjust the pH of the solution to 8.5 using NaOH, and then fill to 100 mL with water (*see* **Note 3**).

2.6 TMT Desalting

1. Materials provided in the set of hydrophobic resin peptide desalting spin columns (*see* **Note 4**).
2. Conditioning/washing solution: 0.1% trifluoroacetic acid.
3. Elution solution: 0.1% trifluoroacetic acid, 50% acetonitrile.
4. TMT washing solution: 5% methanol, 0.1% trifluoroacetic acid.

2.7 LC-MS/MS Analysis

1. High-resolution mass spectrometer (*see* **Note 5**).
2. Data analysis software (*see* **Note 6**).

3 Methods

All methods are performed at room temperature unless implicitly stated otherwise.

3.1 Inoculation and Sample Preparation of T. aestivum with Deoxynivalenol

1. Using 15 cm pots, seed both FHB-resistant and FHB-susceptible cultivars of wheat and grow in a greenhouse or growth chamber at 21/18 °C and 16-h light/8-h dark until the anthesis stage.
2. Upon anthesis, move the plants to a greenhouse or growth chamber with a day/night cycle of 16/8 h at 27/22 °C, respectively, and relative humidity of 75–80%.
3. Point-inoculate 10 μL of the mock, 0.1 mg/mL, and 1.0 mg/mL 15-ADON treatments within the spikelets of a spike at the middle of the head by opening the lemma and palea of the chosen plants.
4. Gently close the spikelet and create an environment of 100% relative humidity by misting the inoculated head and covering with a glycerin bag.
5. Harvest and freeze the heads after 24 and 120 h post-inoculation (hpi), and then store at −80 °C until use.

3.2 Proteome Extraction and Digestion

1. Add proteinase inhibitor cocktail tablet to 10 mL of cold 100 mM Tris–HCl buffer, pH 8.5, and vortex until solubilized.
2. Add inoculated spikelet to a 2 mL low protein binding micro-centrifuge tube with 1 cc of the stainless-steel beads and 350 μL of the freshly prepared proteinase inhibitor solution.
3. Place the tube into the tissue homogenizer to pulverize the samples (*see* **Note 7**).

4. Centrifuge the samples briefly to only collect the material from the sides of the tube but not enough to create a pellet.
5. Transfer 300 μL of sample to a new low protein binding microcentrifuge tube and remove the dirty beads for further cleaning (*see* **Note 8**).
6. Add 1/10 (*v*/*v*) 20% SDS to the sample (final concentration 2% SDS).
7. Add 1/100 (*v*/*v*) 1 M DTT to the sample (final concentration 10 mM DTT).
8. Briefly vortex the solution to mix.
9. Allow the samples to incubate for 10 min at 95 °C while shaking at 800 rpm.
10. Cool the samples to room temperature either slowly acclimating or quickly on ice.
11. Avoiding exposure to light, add 1/10 volume of 0.55 M IAA to the sample (final concentration 55 mM IAA).
12. Briefly vortex the solution to mix.
13. Incubate the solution in the dark at room temperature for 20 min.
14. To the sample, add 100% ice-cold acetone to a final concentration of 80%.
15. Store samples at −20 °C overnight.
16. Centrifuge samples at 4 °C for 10 min at 13,500 rpm to form a pellet.
17. Discard the supernatant and wash the pellet with 80% ice-cold acetone.
18. Repeat **steps 16–17** twice.
19. Discard the supernatant.
20. Allow any remaining acetone to air-dry from the pellet.
21. Reconstitute the pellet in 100 μL 8 M urea/40 mM HEPES.
22. Sonicate in a 4 °C water bath for 5–7 cycles (1 min on/1 min off) or vortex to dissolve the pellet.
23. Measure the concentration of protein in the sample using a desired spectrophotometry protocol (e.g., BSA method).
24. Add 3/4 *v*/*v* ammonium bicarbonate to the samples, to a final concentration of 2 M urea.
25. Remove the aliquot required to contain 100 μg of protein from the solution and add to a new LoBind tube.
26. Digest the 100 μg of protein with 1:50 (enzyme to protein) of the trypsin/Lys-C mixture.
27. Tap to mix through the solution and incubate at room temperature overnight.

3.3 Preparing C18 STAGE Tips and Desalting Samples

1. Quench the digestion using 1/10 stopping solution (*v*/*v*) and centrifuge to create a pellet at 13,500× *g* for 10 min. Remove supernatant and add to a new low protein binding microcentrifuge tube.
2. Create the STAGE tips by inserting three layers of C18 resin into the end of a 200 μL micropipette tip.
3. Wash the resin by adding 100 μL of anhydrous acetonitrile to the tip and remove the liquid by centrifuging at 1000× *g*.
4. Add 50 μL of buffer B to the tip and remove by spinning at 1000× *g*.
5. Add 200 μL of buffer A to the tip and remove by spinning at 1000× *g*.
6. Load 200 μL of the sample onto the STAGE tip and centrifuge at 1000× *g* until liquid has passed through. If the sample is larger than 200 μL, load only 200 μL at a time.
7. Wash the tip with 200 μL of buffer A and spin at 1000× *g* until no liquid remains in the tip.
8. Slowly elute the peptides from the C18 resin with 50 μL buffer B using a syringe into 0.2 mL PCR strip tubes.
9. Dry in a vacuum centrifuge at 45 °C (typically takes 30–40 min).
10. Store dried peptides at −20 °C until ready for either TMT labeling or LC-MS/MS if performing label-free analysis (in the latter, proceed directly to Subheading 3.5).

3.4 TMT Labeling

1. Label samples as per the protocol outlined in the TMT labeling kit (*see* **Note 9**).
2. Desalt the samples as per the protocol outlined in the hydrophobic resin peptide desalting spin columns.
3. Dry the TMT-labeled peptides in a vacuum centrifuge at 45 °C (time varies depending on sample volume).

3.5 LC-MS/MS Analysis

1. Resuspend the dried peptides using 10 μL buffer A. Load sample amount as per mass spectrometer guidelines and user preference. For example, we typically run TMT-labeled complex plant samples on a 3-h gradient using a high-resolution mass spectrometer.

3.6 Proteome Data Analysis

1. Import the raw spectra data into desired peptide analysis software [16].
2. The settings to analyze the raw data are as default, except for the following settings: reporter ion MS2 or MS3 level quantification (*see* **Note 10**), loading the information required to analyze the mass shifts due to the 16plex TMT labels, minimum peptides required to identify a protein to two, and match between runs with a match time window of 0.7 min. Download

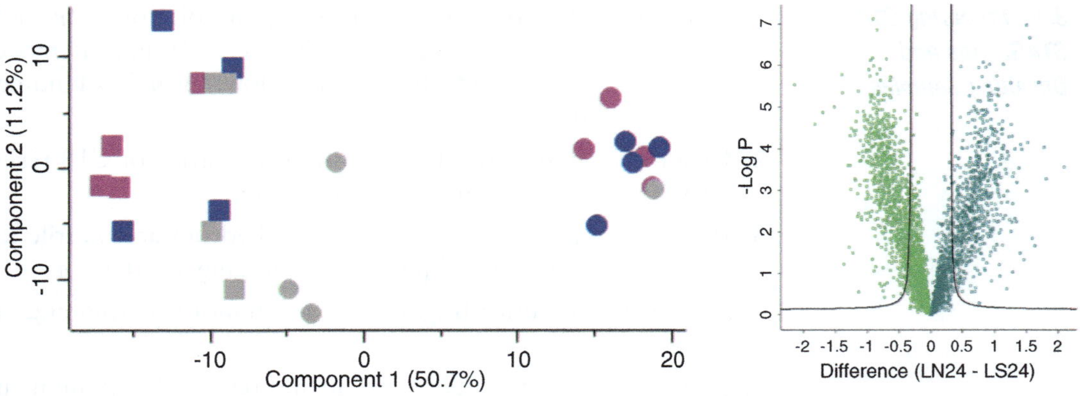

Fig. 2 Example proteomic profiling dataset of FHB-resistant vs. FHB-susceptible cultivars inoculated with DON. (**a**) Principal component analysis (PCA) plot of samples inoculated for 24 h. The FHB-susceptible cultivar samples are shown in squares, while FHB-resistant samples in circles. Mock inoculated (grey) samples, 0.1 mg/mL DON (blue), and 1.0 mg/mL DON (pink). (**c**) Volcano plot comparing difference in protein abundance between FHB-resistant (green) and FHB-susceptible (blue) cultivars inoculated with 0.1 mg/mL DON for 24 hpi; Difference (log_2), -log_{10} p-value reported

the proteome FASTA file for *Triticum aestivum* from UniProt and import it to the MaxQuant software, changing the identifier rule to the UniProt identifier.

3. For data analysis and visualization, label the samples and filter potential contaminants and reverse peptide sequences [17]. Complete data analysis as required. Examples of data generated from a pilot study are provided (Fig. 2).

4 Notes

1. We use the FHB-resistant Sumai#3 and FHB-susceptible Norwell cultivars.
2. We use the Thermo Scientific™ TMTpro™ 16plex Label Reagent Set.
3. The 100 mM HEPES solution could be replaced with a 100 mM TEAB solution.
4. We use the Thermo Scientific™ Pierce™ Peptide Desalting Spin Columns.
5. We recommend a high-resolution mass spectrometer coupled with nano-flow liquid chromatography with a *m/z* range of approx. 50–6000. For example, a mass spectrometer capable of performing MS^3 is valuable for reducing ratio compression; however, a high-resolution mass spectrometer with ion mobility (e.g., Tims-TOF Pro, Exploris 240 or 480 with FAIMS) is also applicable.
6. We use MaxQuant for spectral analysis and Perseus for data processing and visualizations.

7. Pulverize samples at power 8 for 1 min, then remove samples, and place on ice until cool. Return samples to the tissue homogenizer and run at power 10 for 1 min.
8. Place beads into a separate beaker for washing. Rinse with water twice, and then submerge in ethanol. Allow the ethanol to evaporate in a fume hood, and then rinse again with dH_2O. Autoclave beads and store for future use.
9. Create a common channel consisting of all samples to an equivalent protein concentration as the individually labelled samples. This allows for comparative analysis between runs when dealing with large amounts of samples.
10. Reporter ion-level quantification is dependent upon the mass spectrometer used for the LC-MS/MS analysis.

References

1. Figueroa M, Hammond-Kosack KE, Solomon SE (2018) A review of wheat diseases – a field perspective. Mol Plant Pathol 19:1523–1536
2. Desjardins AE, Proctor RH (2007) Molecular biology of *Fusarium* mycotoxins. Int J Food Microbiol 119:47–50
3. Boenisch MJ, Schäfer W (2011) *Fusarium graminearum* forms mycotoxin producing infection structures on wheat. BMC Plant Biol 11:110
4. Mccormick SP, Stanley AM, Stover NA et al (2011) Trichothecenes: from simple to complex mycotoxins. Toxins 3:802–814
5. Eriksen GS, Pettersson H (2004) Toxicological evaluation of trichothecenes in animal feed. Anim Feed Sci Tech 114:205–239
6. Gunnaiah R, Kushalappa AC, Duggavathi R et al (2012) Integrated metabolo-proteomic approach to decipher the mechanisms by which wheat QTL (Fhb1) contributes to resistance against *Fusarium graminearum*. PLoS One 7:e40695
7. Walter S, Doohan F (2011) Transcript profiling of the phytotoxic response of wheat to the *Fusarium* mycotoxin deoxynivalenol. Mycotoxin Res 27:221–230
8. Perochan A, Jianguang J, Kahla A et al (2015) *TaFROG* encodes a *Pooideae* orphan protein that interacts with SnRK1 and enhances resistance to the mycotoxigenic fungus *Fusarium graminearum*. Plant Physiol 169:2895–2906
9. Geddes J, Eudes F, Laroche A et al (2008) Differential expression of proteins in response to the interaction between the pathogen *Fusarium graminearum* and its host, *Hordeum vulgare*. Proteomics 8:545–554
10. Ding L, Li M, Li P et al (2017) Comparative proteomics analysis of young spikes of wheat in response to *Fusarium graminearum* infection. Acta Physiol Plant 39:271
11. Fabre F, Urbach S, Roche S et al (2021) Proteomics-based data integration of wheat cultivars facing *Fusarium graminearum* strains revealed a core-responsive pattern controlling Fusarium head blight. Front Plant Sci 12:644810
12. Eldakak M, Das A, Zhuang Y et al (2018) A quantitative proteomics view on the function of *Qfhb1*, a major QTL for Fusarium head blight resistance in wheat. Pathogens 7:58
13. Liu J, Li L, Foroud NA et al (2019) Proteomics of bulked rachides combined with documented QTL uncovers genotype nonspecific players of the Fusarium head blight responses in wheat. Phytopathology 109:111–119
14. Zecha J, Satpathy S, Kanashova T et al (2019) TMT labeling for the masses: a robust and cost-efficient, in-solution labeling approach. Mol Cell Proteomics 18:1468–1478
15. Thompson A, Wölmer N, Koncarevic S et al (2019) TMTpro: design, synthesis, and initial evaluation of a proline-based isobaric 16-plex tandem mass tag reagent set. Anal Chem 91:15941–15950
16. Cox J, Mann M (2008) MaxQuant enables high peptide identification rates, individualized p.p.b.-range mass accuracies and proteome-wide protein quantification. Nat Biotechnol 26:1367–1372
17. Tyanova S, Temu T, Sinitcyn P et al (2016) The Perseus computational platform for comprehensive analysis of (prote)omics data. Nat Methods 13:731–740

Chapter 13

Quantitative Phosphoproteome Analysis of the Interaction Between *Fusarium graminearum* and *Triticum aestivum*

Boyan Liu, Danisha Johal, Reid Buchanan, Brianna Ball, Mitra Serajazari, and Jennifer Geddes-McAlister

Abstract

Quantitative proteomics is a powerful method for distinguishing protein abundance changes in a biological system across conditions. In addition to recent advances in computational power and bioinformatics methods, improvements to sensitivity and resolution of mass spectrometry (MS) instrumentation provide an innovative approach for studying host-pathogen interaction dynamics and posttranslational modifications. In this protocol, we provide a workflow for state-of-the-art MS-based proteomics to assess changes in phosphorylated protein abundance upon interaction between the worldwide cereal crop, *Triticum aestivum* (wheat), and the global cereal crop fungal pathogen, *Fusarium graminearum*, during infection. This protocol mimics a time course of infection of *T. aestivum* by *F. graminearum* in the greenhouse, and the harvested samples undergo Fe-NTA phosphoenrichment combined with label-free quantification (LFQ) for detection by liquid-chromatography (LC)—coupled with tandem MS/MS. Our approach provides an in-depth view of changes in phosphorylation from both the host and pathogen perspectives in a single experiment across infection time points and different host cultivars.

Key words *Fusarium graminearum*, Phosphoproteome, Quantitative proteomics, *Triticum aestivum*, Label-free quantification (LFQ), Mass spectrometry (MS), Host-pathogen interaction, Bioinformatics

1 Introduction

Fusarium graminearum causes the worldwide cereal crop disease, Fusarium head blight (FHB), which leads to millions of dollars in losses annually through fungal contamination and mycotoxin accumulation [1]. The primary mycotoxin produced by *F. graminearum* is deoxynivalenol (DON), which accumulates in crops, contaminating feed and water supplies, threatening the livestock and poultry industries, and, consequentially, affecting human health [2]. Current methods to constrain fungal infection in the field rely on seeding of FHB-resistant cultivars that still display some susceptibility depending on the environmental conditions

Nora A. Foroud and Jonathan A. D. Neilson (eds.), *Plant-Pathogen Interactions*, Methods in Molecular Biology, vol. 2659,
https://doi.org/10.1007/978-1-0716-3159-1_13,

and the fungal strain present, along with the administration of a fungicide at heading, which can promote the evolution of resistance. Moreover, it is predicted that fungal outbreaks will become more severe over time due to climate change, supporting the need to develop new resistant varieties and/or the discovery of biocontrol agents [3].

To combat FHB, the interaction between the host and pathogen requires extensive profiling to comprehensively understand how the two biological systems defend against one another and to uncover new strategies for overcoming infection. Recent advances in mass spectrometry (MS)-based proteomics enable studying host-pathogen interactions on a larger scale than before. By not limiting analysis to the cellular proteome, but expanding to the secretome, interactome, and investigation of posttranslational modifications, comprehensive information about cellular regulation and biological processes is acquired [4]. One of the most important pieces in host-pathogen interactions is the communication and signaling events that occur between and within the biological systems, including phosphorylation events, which regulate protein abundance, activity, and function during infection. To date, the phosphoproteome between *F. graminearum* and its host has been defined under in vitro conditions or with a focus on a single perspective (i.e., either host or pathogen), but have yet to be fully characterized from dual perspectives on a large scale [5].

In this protocol, we identify phosphoproteins associated with the temporal infection process via phosphopeptide enrichment on *F. graminearum*-infected *Triticum aestivum* (wheat) samples with both FHB-resistant and FHB-susceptible cultivars (Fig. 1). Overall, by investigating phosphorylation modifications with MS-based proteomics, this protocol provides a robust and transferrable outline to identify fungal signaling events of environmental adaptation strategies in plant-associated fungal pathogens and uncover new options to combat disease.

2 Materials

UltraPure water (18 MΩ at 25 °C) and analytical grade reagents are used to prepare all solutions. All steps are performed at room temperature unless otherwise specified. All waste disposal regulations are followed during disposition of waste materials. Solutions are prepared fresh unless indicated otherwise.

2.1 Media for Inoculation

1. *Fusarium graminearum* strain PH-1 (or desired strain).
2. Potato dextrose agar plates.
3. Wheat straw media: Autoclave 5 g of ground wheat straw with 125 mL water in a 250 mL Erlenmeyer flask.

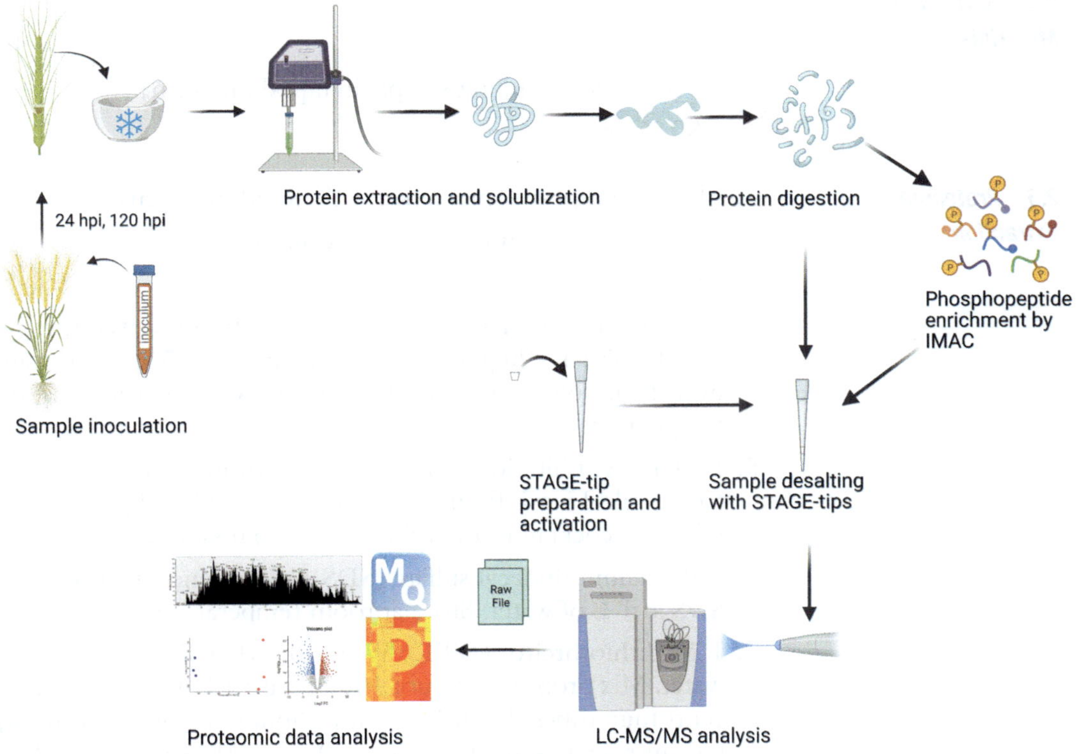

Fig. 1 Workflow of the mass spectrometry-based proteomics protocol for measuring and quantifying changes in protein abundance during Fusarium head blight infection of wheat. Image produced by biorender.com

4. Rotary shaker.
5. Hemocytometer and hemocytometer coverslip (20 mm × 26 mm × 0.5 mm thick).
6. Sterile water.

2.2 Plant Materials

1. *Triticum aestivum* seeds (resistant and susceptible cultivars).
2. Clean pots (e.g., 15 cm pots).
3. Soil (autoclaved).
4. Greenhouse or growth room (with appropriate temperature and lighting; e.g., 21/18 °C with 16/8 h photoperiod).

2.3 Inoculation of T. aestivum

1. Inoculum: Wash spores with sterile water and dilute to 4000 macroconidia/μL with sterile water.
2. Mock inoculum: Sterile water.
3. Marker or string.
4. 10 μL pipette and sterile pipette tips.
5. Misting bottle.
6. Plastic bags.

2.4 Harvesting Materials

1. Aluminum bag.
2. Liquid nitrogen.
3. Tweezer (for handling sample in liquid nitrogen).
4. −80 °C freezer.

2.5 Proteome Extraction

1. Mortar and pestle (prechilled with liquid nitrogen).
2. Probe sonicator and water bath sonicator.
3. Thermal shaker.
4. Tris–HCl buffer (100 mM, pH 8.5): To 900 mL water, weigh and dissolve 10.90 g Tris. Adjust the pH to 8.5 and add water until final volume reaches 1 L. Autoclave and store at room temperature.
5. Proteinase inhibition buffer: Transfer 10 mL Tris–HCl buffer to a sterile 15 mL falcon tube and add 1 tablet of phosphatase inhibitor cocktail to the tube. Vortex to dissolve.
6. 20% sodium dodecyl sulfate (SDS): Weigh and dissolve 200 g SDS in 1 L of water. Store at room temperature.
7. 1 M dithiothreitol (DTT): Weigh 0.781 g DTT in to 5 mL water. Mix properly. Aliquot 100 μL into 1.5 mL sterile microcentrifuge tubes. Flash-freeze the aliquoted reagents with liquid nitrogen and store at −20 °C. All steps should be performed in the fume hood. Discard any unused reagent after thawing.
8. 0.55 M iodoacetamide (IAA): Weigh 2.03 g IAA in to 20 mL water. Protect from light by wrapping with aluminum foil. Mix properly until all solids dissolve. Aliquot 500 μL into sterile 1.5 mL microcentrifuge tubes in the fume hood. Flash-freeze the aliquoted reagents with liquid nitrogen and store at −20 °C. All steps should be performed in the fume hood. Discard any unused reagent after thawing.
9. 100% and 80% acetone: Store in −20 °C freezer.
10. 8 M urea/40 mM HEPES: Weigh and dissolve 0.19 g HEPES and 9.6 g urea in 20 mL water. Flash-freeze the aliquoted reagents with liquid nitrogen and store at −20 °C. Discard of any unused reagent after thawing.
11. 50 mM ammonium bicarbonate (ABC): Weigh and dissolve 0.40 g of NH_4HCO_3 in 100 mL water. Mix well and store at room temperature.
12. Lys-C/trypsin mixture: Reconstitute according to the manufacturer's instructions.
13. 15 mL conical tube.
14. 2 mL LoBind tubes.

2.6 Phosphopeptide Enrichment by Immobilized Affinity Chromatography (IMAC)

1. Phosphopeptide enrichment kit (*see* **Note 1**).
2. Vacuum centrifuge.

2.7 Stop-and-Go Extraction Tips (STAGE-Tip) Desalting

1. 10 mL syringe (without the needle).
2. PCR strip tubes and caps.
3. 200 μL sterilized pipette tips.
4. Stopping solution: 20% acetonitrile (ACN), 6% trifluoroacetic acid (TFA).
5. 100% ACN: Store at room temperature.
6. Buffer A: 2% ACN, 0.1% TFA, 0.5% acetic acid (all in *v*/*v*).
7. Buffer B: 80% ACN, 0.5% acetic acid (all in *v*/*v*).
8. C_{18} resin.

2.8 Mass Spectrometry

1. High-resolution mass spectrometer (*see* **Note 2**).
2. Data analysis software (*see* **Note 3**).

3 Methods

All steps are performed at room temperature unless otherwise specified.

***3.1 Culturing* F. graminearum *and Inoculating* T. aestivum**

1. Seed FHB-susceptible and FHB-resistant wheat (*see* **Note 4**) genotypes in 15 cm pots at 21/18 °C with 16/8 h photoperiod until the anthesis stage in the growth room. Water every other day, as needed.
2. Culture the desired strain of *F. graminearum* onto potato dextrose agar (PDA) plate for 5 d at room temperature and dark conditions.
3. Produce a macroconidia suspension by transferring four PDA plugs (*see* **Note 5**) to the autoclaved water + wheat straw media and incubate the culture on a rotary shaker (120 rpm) at 25 °C for 2 weeks.
4. For inoculum, count the macroconidia using a hemocytometer, collect and wash with sterile water, and resuspend to 4000 macroconidia/μL in sterile water. Use sterile water for the mock inoculum.
5. Pipet 10 μL of macroconidia or mock inoculum into the targeted spikelet (*see* **Note 6**).

6. Mist the newly inoculated heads with sterile water (*see* **Note 7**), and then cover with a plastic bag. Place the plants inside a growth chamber with 27/22 °C with 16/8 h photoperiod.
7. Harvest wheat heads after 24 and 120 h post inoculation (hpi) and immediately flash-freeze in aluminum bags with liquid nitrogen. Samples are stored in a −80 °C freezer until proteome extraction.

3.2 Protein Extraction and Solubilization

1. Collect 12–18 inoculated or mock-inoculated spikelets (*see* **Note 8**, Fig. 2) and grind into a fine powder using a precooled mortar and pestle in the presence of liquid nitrogen (*see* **Note 9**).
2. Transfer the powder to a 15 mL tube (*see* **Note 10**) and add 2 mL of cold 100 mM Tris–HCl (contains phosphoproteinase inhibitor cocktail tablet) immediately.
3. Add 20% SDS to final concentration of 2% (1/10, SDS to sample volume) and 1:100 volume of 1 M stock DTT (final concentration of 10 mM DTT) directly to the 15 mL tube.
4. Sonicate at 30 s on/30 s off for 2–5 cycles using a probe sonicator (120 Watt). Briefly vortex the samples, and incubate on thermal shaker at 98 °C for 10 mins with 800 rpm rotation.
5. Cool samples to room temperature (*see* **Note 11**) before adding 1:10 volume of 55 mM IAA to the samples (final concentration of 5 mM IAA). Incubate samples for 20 mins in the dark at room temperature to avoid light exposure.
6. Remove plant debris by filtering through a 40 μm pore-size filter (*see* **Note 12**) and transfer the supernatant into a clean 2 mL LoBind tube.
7. Centrifuge samples at 10,000 × *g* at 4 °C for 10 min to pellet any remaining debris and carefully transfer the supernatant to a new tube.
8. Add 100% ice-cold acetone to the sample (final concentration of acetone at 80%) and incubate at −20 °C overnight.

3.3 Protein Quantification and Digestion

1. Centrifuge samples at 10,000 × *g* and 4 °C for 10 min to pellet precipitate.
2. Discard supernatant and transfer the pellet into 2 mL LoBind tubes, and wash the pellet with 1 mL of 80% ice-cold acetone.
3. Centrifuge the sample at 10,000 × *g* and 4 °C for 10 min, discard the supernatant, and repeat **step 2** again.
4. Dry the pellet completely (*see* **Note 13**).
5. Resuspend the pellet in 400 μL of 8 M urea/40 mM HEPES (*see* **Note 14**). Sonicate the samples in a 4 °C water bath for 5–7 cycles (1 min on/1 min off) (*see* **Note 15**).

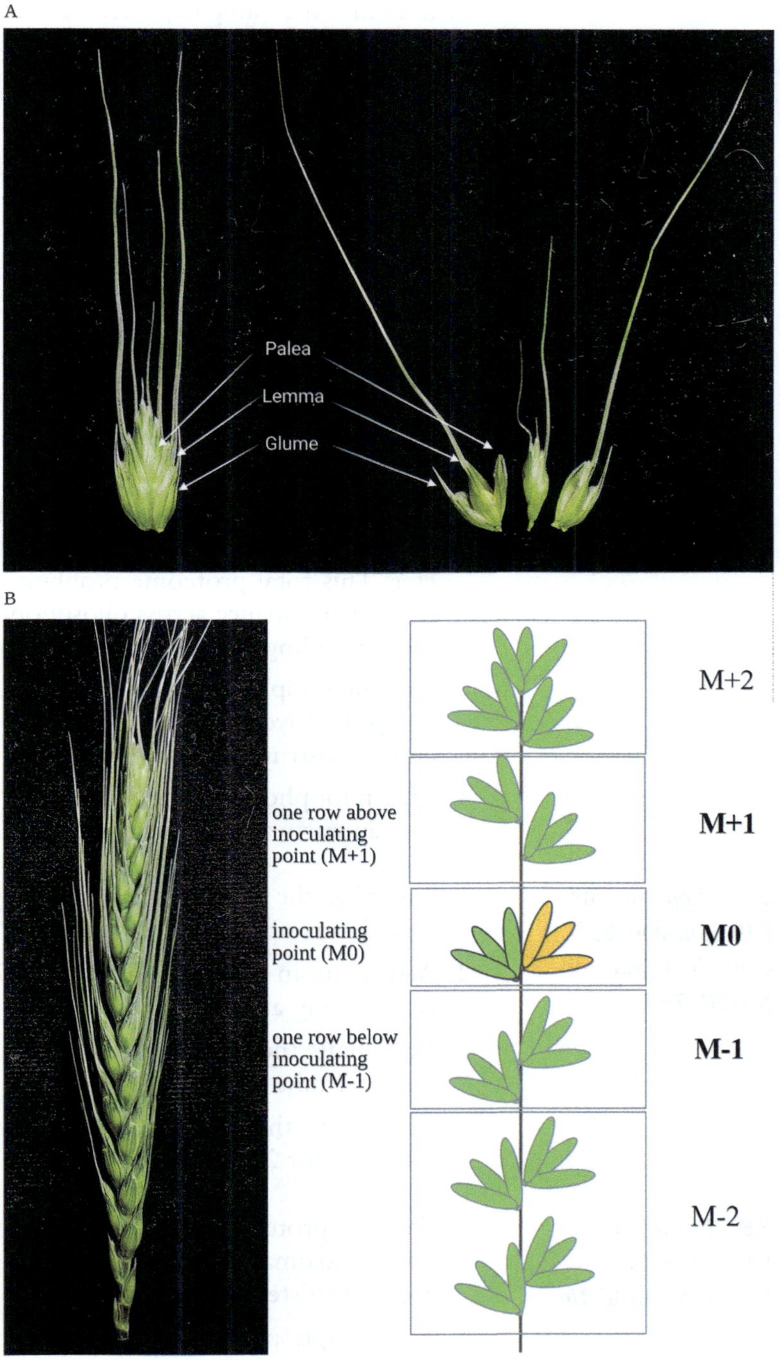

Fig. 2 Inoculation and collection of wheat with Fusarium head blight. (**a**) Structure of wheat head. Point inoculation site is between lemma and palea. (**b**) The point inoculation site on wheat head. M0 represents the inoculated spike. M + 1 is the row above the infected site; M-1 is the row below the infected site

6. Measure the sample protein concentration using a protein quantification assay (e.g., BSA urea assay) (*see* **Note 16**).
7. Add 50 mM ammonium bicarbonate to adjust the final urea concentration in the sample to 2 M.
8. Transfer 600–800 μg of protein (or the amount of protein needed according to the phosphorylation enrichment kit) from the sample tube to a new 2 mL LoBind tube for digestion.
9. Add Lys-C and trypsin protease mixture to ~800 μg of protein in 1:50 enzyme-to-protein ratio (μg). Digest at room temperature overnight.

3.4 Phosphopeptide Enrichment by IMAC

1. Add 1/10 volume of stopping solution and centrifuge at 16,000 × *g* to collect the supernatant in a new 2 mL LoBind tube.
2. Collect 1/10 volume of digested sample (approximately 50–100 μg) for total proteome profiling on the mass spectrometer. This total proteome profiling can be used to normalize protein abundance across phosphoproteome samples. Proceed to Subheading 3.5 with the total proteome aliquot.
3. Dry the sample in a vacuum centrifuge until the peptides are completely lyophilized. Depending on the amount of liquid and the instrument, this process may take up to 2 h (at 45 °C).
4. Enrich for phosphopeptides using an enrichment kit according to manufacturer's instructions (*see* **Notes 1** and **17**).

3.5 Preparing the C18 Stop-and-Go Extraction Tips (STAGE-Tip)

1. Assemble the STAGE-tip in a 200 μL micropipette tip with three layers of C18 resin.
2. Activate resin with 100 μL 100% ACN. Remove the liquid by centrifuging at 1000 × *g* for 2 min.
3. Equilibrate the tip with 50 μL buffer B by centrifuging at 1000 × *g* for 2 min.
4. Equilibrate the tip with 200 μL buffer A by centrifuging at 1000 × *g* for 2 min.

3.6 Desalting Samples with Prepared STAGE-Tip

1. For total proteome sample collected in Subheading 3.4, **step 1**, add approximately 50–100 μg of sample to the STAGE-tip and proceed to **step 4** (*see* **Note 18**).
2. For phosphoproteome samples collected in Subheading 3.4, **step 4**, resuspend the lyophilized phosphopeptide samples in 200 μL of buffer A (*see* **Note 19**).
3. Sonicate dissolved samples in water bath at 4 °C to ensure pellet has dissolved before loading to the STAGE-tip.

4. Centrifuge the loaded STAGE-tip at 1000 × *g* for 3–5 min, or until the sample has moved through the tip (*see* **Note 20**).
5. Wash the tip with 200 μL buffer A. Centrifuge at 1000 × *g* for 2 min to remove the liquid.
6. Elute sample from the tip with 50 μL buffer B using a syringe. Collect the eluted sample in a 0.2 mL PCR strip tube (*see* **Note 21**).
7. Lyophilize eluted peptides in a vacuum centrifuge (45 °C for 30–40 min).

3.7 LC-MS/MS Analysis

1. Reconstitute peptides in 10 μL buffer A and load the required amount of peptides onto a high-resolution mass spectrometer (*see* **Note 2**), depending on instrument specifications and user preference.

3.8 Phosphoproteome (and Total Proteome) Data Analysis

1. Analyze the obtained MS spectra on the desired platform (*see* **Note 3**).
2. For spectra analysis of both total proteome and phosphoproteome, the unprocessed files from the mass spectrometer are loaded into the alignment and identification software (*see* **Note 22**). Parameters are set for quantification (e.g., LFQ) and modifications, including phosphorylation at STY sites (*see* **Note 23**). Digestion enzymes are selected according to the experiment. Load FASTA files for *Fusarium graminearum* and *Triticum aestivum* from UniProt for peptide identification. Set the false discovery rate (FDR) for peptide level and protein group level to 1%. Start run.
3. For data analysis, load protein groups' output file into the software. Filter for contaminants, reverse peptides, and peptides only modified by site (for the total proteome files). For phosphoproteome files, filter rows based on numerical column with localization probability of 1 and number of relations as $x > 0.75$. Next, $\log_2$ transform the datasets, add categorical annotations to each column, filter for valid values (>50%), add imputation values based on the normal distribution (as appropriate), and process the data, as per user preference. A representative sample of the phosphoproteome data set is provided (Fig. 3).

4 Notes

1. For phosphopeptide enrichment, we use the High-Select Fe-NTA phosphopeptide enrichment kit (Thermo Fisher Scientific), but alternative kits would also be effective.

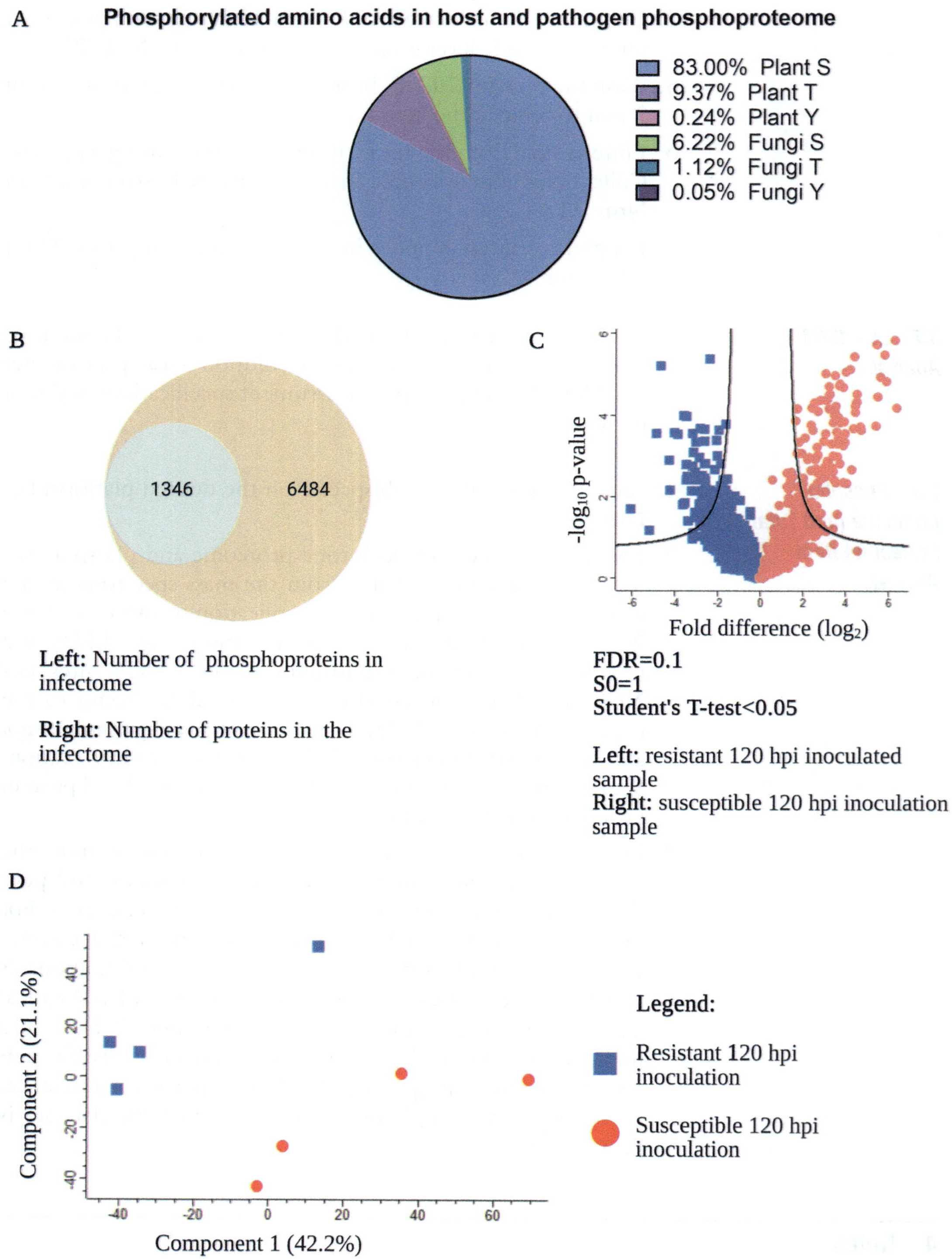

Fig. 3 Sample phosphoproteome dataset for profiling Fusarium head blight infection of wheat from dual perspectives. (**a**) Pie chart of phosphorylated peptides in host and pathogen. (**b**) Venn diagram showing the number of phosphoproteins in the infectome (left) and total proteins in the infectome (right). (**c**) Volcano plot of phosphorylated peptides. Left, resistant (e.g., Sumai #3) 120 hpi *F. graminearum*-inoculated sample; right, susceptible (e.g., Norwell) 120 hpi *F. graminearum*-inoculated sample. Student's *t*-test p-value <0.05; FDR $= 0.05$; S0 $= 1$. (**d**) Principal component analysis of infectome

2. We use a Thermo Scientific Orbitrap Exploris 240. A high-resolution mass spectrometer coupled with nano-flow liquid chromatography with a *m/z* range of approximately 50–6000 is recommended.
3. We use MaxQuant [6] for spectra analysis and Perseus for data processing and visualization [7].
4. We use Norwell and Sumai#3 as an FHB-susceptible and a resistant cultivar, respectively.
5. Select the plug with more aerial mycelia and use sterile straws and sticks to take out plugs, or simply do it with the bottom of pipette tips.
6. Mix the inoculum properly before inoculating. Otherwise, the macroconidia may not distribute evenly in the suspension. Mark the inoculated floret with a marker or a string, or simply attach a string to it to distinguish the inoculation site from the rest. This is especially important for the 24 hpi samples as they have relatively weaker symptom than at 120 hpi.
7. Distance the spray head from the infected head, or the macroconidia may be washed off. Cover the infected head with a plastic bag to ensure high humidity during infection.
8. Keep the samples on ice for this section whenever possible.
9. Clean the mortar and pestle with 70% ethanol and sterile water between the biological replicates. Wash with detergent, water, and 70% ethanol in between the samples with different biological conditions (e.g., wash between 120 hpi susceptible and resistant samples).
10. Keep mortar as cold as possible to ensure all powders are transferred. Keep the 15 mL tube cold, if possible, during the process. You can also use a liquid nitrogen-chilled spatula to assist the transfer process.
11. Place the samples on ice for fast cooling. We found no difference between fast-cooling and slow-cooling sample (cool at room temperature).
12. Pore sizes ranging from 40 μm to 70 μm can be used. The single-use cell strainer is a good option here. Pour the sample into a cell strainer and centrifuging the debris at the lowest speed for less than 10 s is sufficient to collect all liquid from the solid debris.
13. Leave the open tubes in a biosafety cabinet at room temperature for 30 min, or heat the tubes at 37 °C with the lid open to dry the sample quickly.
14. Adjust the volume of urea/HEPES according to the pellet size to ensure all pellets are completely dissolved. For 18 spikes, about 400 μL of 8 M ureas/40 mM HEPES can be used. Lengthen the sonication cycle if needed.

15. Adjust the sonication period according to the pellet size. A longer sonication period (e.g., 3 min on/3 min off for six cycles) is recommended for large pellet.
16. For phosphopeptide enrichment, between 500 μg and 1000 μg of protein is recommended.
17. Phosphorylated samples cannot be stored in the elution buffer. Dry samples immediately after elution with a vacuum centrifuge to avoid loss of phosphorylated samples due to high pH elution buffer.
18. Flash-freeze any remaining sample in liquid nitrogen and store at −20 °C for up to 1 week.
19. Adjust the volume of buffer A according to the pellet size to dissolve properly.
20. Centrifugation may take up to 30 min and it is better to centrifuge at room temperature than cold, as precipitate may clog up the STAGE-tips.
21. Use a syringe without the needle and pressure buffer B slowly through the filter. To ensure all samples in the filter are collected, elute 25 μL first followed by another 25 μL, or elute with 50 μL but at a slower elution speed.
22. The total proteome and phosphoproteome MS files can be processed separately or combined to normalize protein abundance due to phosphorylation events.
23. For total proteome analysis, the phospho (STY) modification is not included, but all other parameters are equivalent. Variable modifications include oxidation (methionine) and N-terminal acetylation. Fixed modifications include carbamidomethyl (C).

References

1. Xia R, Schaafsma A, Wu F, Hooker D (2020) Impact of the improvements in Fusarium head blight and agronomic management on economics of winter wheat. World Mycotoxin J 13:423–439
2. Tamburic-Ilincic L, Wragg A, Schaafsma A (2015) Mycotoxin accumulation and Fusarium graminearum chemotype diversity in winter wheat grown in southwestern Ontario. Can J Plant Sci 95:931–938
3. Torres AM, Palacios SA, Yerkovich N et al (2019) Fusarium head blight and mycotoxins in wheat: prevention and control strategies across the food chain. World Mycotoxin J 12: 333–355
4. Ball B, Sukumaran A, Geddes-McAlister J (2020) Label-free quantitative proteomics workflow for discovery-driven host-pathogen interactions. J Vis Exp e61881
5. Yun Y, Liu Z, Yin Y et al (2015) Functional analysis of the *Fusarium graminearum* phosphatome. New Phytol 207:119–134
6. Cox J, Mann M (2008) MaxQuant enables high peptide identification rates, individualized P.P. b.-range mass accuracies and proteome-wide protein quantification. Nat Biotechnol 26: 1367–1372
7. Tyanova S, Temu T, Sinitcyn P et al (2016) The Perseus computational platform for comprehensive analysis of (prote)omics data. Nature Methods 13:731–740

Chapter 14

Fatty Acid Profiling of Grapevine Extracellular Compartment

Ana Rita Cavaco, Gonçalo Laureano, Andreia Figueiredo, and Ana Rita Matos

Abstract

The apoplast is the plant compartment present between the plasma membrane and the cuticle, comprised of the cell wall and the extracellular spaces where the "secretomes" are released and where the apoplastic fluid circulates. Within the many functions attributed to this compartment, its role in plant-pathogen interactions is irrefutable. It is the major point where plant and pathogen secretomes come in contact and several plant and pathogenic secreted proteins and small molecules present in this compartment are already cataloged in the literature. In plant-pathogen interactions, fatty acids and lipid molecules were shown to play a crucial role in the activation of plant immunity; however, the lipid composition of the apoplast is still a black box. Most of the studies performed to understand apoplast dynamics have used proteomic-based techniques; however, knowledge about apoplastic proteins involved in lipid metabolism and transport is still severely limited. In grapevine, only three studies have been published so far focusing on the characterization of this compartment and only one of them deals with grapevine-pathogen interaction. Here we refer to our recently established method for grapevine leaves' apoplastic fluid isolation and describe a direct methylation protocol for the analysis of apoplastic fluid fatty acids. We also point out a novel intracellular marker that may be used to assess apoplastic fluid purity.

Key words Fatty acid profiling, Direct methylation, Gas chromatography, Apoplast, Apoplastic fluid, Grapevine leaves, C16:1 *t*, *Vitis vinifera*

1 Introduction

Grapevine (*Vitis vinifera* L.) is one of the most economically important crops worldwide, mainly due to its use for wine and table grape production. However, grapevine is prone to several diseases. Downy and powdery mildews and grey mold, caused by *Plasmopara viticola*, *Erysiphe necator*, and *Botrytis cinerea*, respectively, are among the most devastating [1]. Disease control strategies include phytochemical applications every growing season, which compromise viticulture sustainability. Understanding the molecular processes behind disease resistance or susceptibility is vital to define alternative control strategies and select new disease

Nora A. Foroud and Jonathan A. D. Neilson (eds.), *Plant-Pathogen Interactions*, Methods in Molecular Biology, vol. 2659,
https://doi.org/10.1007/978-1-0716-3159-1_14,

resistance traits for breeding programs. Lipids and lipid-derived metabolites are not only major structural constituents of cells, but they also function as modulators of a multitude of signal transduction pathways evoked by biotic stresses [2]. Fatty acids (FA) are the main component of cell membranes and an important source of energy. Besides their structural and metabolic roles, FA and FA metabolites participate in plant defense, modulating a myriad of signal transduction pathways. The polyunsaturated FAs (PUFAs) linoleic and linolenic acids are the major constituents of membrane lipids, and under pathogen attack, they can be released by phospholipases A, acting directly in plant defense as free FA or indirectly by being the substrates for oxylipin production [3]. The modulation of membrane lipid composition, namely, the degree of fatty acid unsaturation, in response to pathogen attack, can also impact membrane fluidity and consequently stress response [4, 5]. Moreover, specific fatty acids were shown to be involved in plant resistance against pathogens with different colonization strategies (biotroph, hemibiotroph, and necrotrophy; [5]).

When studying the first moments of plant-pathogen interactions, the apoplast is one of the most important cellular compartments since it is where molecular effectors that trigger the host immune system are secreted by the pathogen [6]. Modulation of apoplast protein composition has been recently started to be unveiled [7]; however, knowledge on apoplastic lipids and their carriers has been severely limited.

The apoplast being the first battlefield where pathogen recognition occurs and secretion of both defense molecules and pathogen effectors take place, understanding its dynamics is crucial. We have previously published a grapevine leaf apoplastic fluid (APF) extraction method compatible with proteomic and metabolomic studies [8]. Here we describe a direct methylation protocol for the analysis of APF fatty acids. We also point out a novel intracellular marker that may be used to assess APF purity.

2 Materials

Prepare all solutions using UltraPure water (prepared by purifying deionized water, to attain a sensitivity of 18 MΩ-cm at 25 °C) and analytical grade reagents. Prepare and store all reagents at room temperature (unless otherwise indicated). Diligently follow all waste disposal regulations.

2.1 Plant Material and Apoplastic Fluid Isolation Buffer

1. Field- or greenhouse-grown grapevine plants of appropriate age and life stage, preferably with fully expanded leaves, should be used (*see* **Note 1**). The protocol is optimized for approximately 25 g of fresh weight of grapevine leaves.

2. Ice-cold infiltration buffer: 100 mM Tris–HCl buffer, pH 7.6. Add about 100 mL water to a 1-L graduated glass beaker, weigh 121.14 g Tris and transfer to the beaker, add water to a volume of 900 mL, and mix with a magnetic stir bar until it dissolves. Adjust pH with HCl. Add 500 mM potassium chloride (37.275 g) and 6 mM CHAPS (3.68 g) and mix until it dissolves. Adjust volume to 1 L with water. Store at 4 °C. Add 2% of sodium sulfite (20 g per L) to the buffer solution just before use [9].
3. Ice-cold deionized water.

2.2 Fatty Acid Profiling Reagents

1. Internal standard: 50 μL 400 μg/mL margaric acid in ethanol/toluene, (1/4, v/v).
2. 3 mL methanol/sulfuric acid (39:1 v/v).
3. 150 μL apoplastic fluid (APF) solution.
4. 3 mL petroleum ether (boiling range 80–100 °C).
5. 2 mL UltraPure water.
6. 10–20 μL hexane.
7. Nitrogen gas.

2.3 Equipment and Consumables

1. Paper towel.
2. Scalpel.
3. Glass beaker.
4. Weighing scale.
5. Large glass Kitasato flask, or vacuum flask, with lid.
6. Vacuum pump to apply vacuum at 25 kPa (187.52 mmHg).
7. 20 mL plastic syringe without plunger.
8. 50 mL Falcon tube or equivalent.
9. 1.5 mL microcentrifuge tubes.
10. Refrigerated high-speed centrifuge with swing rotor for 50 mL centrifuge tubes.
11. Test tube with screw cap (Pyrex culture tubes, 10 mL).
12. p-20, p-100, and p-5000 automatic pipettes.
13. p-20, p-100, and p-5000 pipette tips (no filter).
14. Controllable temperature water bath equipment (temperature range: 37–70 °C).
15. Vortex.
16. Benchtop centrifuge.
17. Glass long-tipped Pasteur pipette.
18. Rubber teat for Pasteur pipettes.
19. Pyrex tubes with PTFE-lined screw cap.

20. Hamilton syringes (10 and 100 μL).
21. Gas chromatograph equipped with a flame ionization detector (FID) and a column for FAMEs. Appropriate software for analysis should be used (*see* **Note 2**).

3 Methods

3.1 Apoplastic Fluid (APF) Extraction

This protocol has been described in detail in [9]. The main steps are as follows:

1. Harvest fully expanded grapevine leaves to obtain approximately 25 g of fresh weight material. Remove the middle and lateral veins using a scalpel.
2. Cut leaves into pieces of around 2 cm^2 to a beaker kept on ice and then transfer to a Kitasato flask filled with an ice-cold infiltration buffer.
3. Apply vacuum to the Kitasato flask employing a 25 kPa around 187.52 mmHg pressure with a vacuum pump (3 min in total).
4. After infiltration, rinse leaf pieces twice in ice-cold deionized water and dry the leaf pieces in soft towel paper.
5. Stack up the infiltrated leaves into a 20 mL syringe and then in a 50 mL centrifuge tube. Inside the centrifuge tube, place a 1.5 mL microcentrifuge tube, with the tip of the syringe oriented to be inside the microcentrifuge tube to collect the APF.
6. Centrifuge at 5000 × g during 15 min at 4 °C to collect the APF to the microcentrifuge tube. Store at −20 °C.

3.2 Fatty Acid Analysis Through Direct Methylation of APF and Leaf Tissue and Gas Chromatography

We have used the direct transesterification of the fresh tissue method for the analysis of the fatty acid composition and contents of grapevine leaves [4, 5, 10] and other high plants and (micro) algae. This method was adapted from an *Arabidopsis thaliana* seed oil quantification method [11]. It has the advantage of being less time-consuming compared with total lipid extraction followed by transesterification of the lipid sample, having a similar or even superior (in the case of some tissues) fatty acid recovery capacity. The water content of the sample should not exceed ~10% of the total reaction volume and the recommended starting material for plant leaves is 50–100 mg, but it is possible to use as little as 10 mg FW.

3.2.1 Preparation of Fatty Acid Methyl Esters (FAMEs) from Grapevine APF and Leaf Samples

1. Program the water bath equipment for 70 °C.
2. Add 50 μL of the internal standard (Subheading 2.2) with a syringe to a Pyrex tube with a PTFE-lined screw cap and dry it under nitrogen atmosphere, until solvent evaporation.

3. Add freshly prepared 3 mL methanol/sulfuric acid (39:1 v/v).
4. Add 150 μL APF sample or 50 mg of homogenized leaf tissue (*see* **Note 3**), screw the cap, vortex briefly, and immediately incubate at 70 °C for 1 h.
5. Stop the reaction by cooling the tubes in room temperature water.
6. Add 3 mL petroleum ether and 2 mL UltraPure water.
7. Vortex and centrifuge the tubes for 5 min at 4000 × *g*.
8. Program the water bath to 37 °C.
9. Recover the upper (organic) phase containing the purified FAMEs with a Pasteur pipette to a test tube and dry it at 37 ° C under nitrogen atmosphere (*see* **Note 4**).
10. Resuspend the sample in 10–20 μL hexane.

3.2.2 Analysis of FAMEs by Gas Chromatography

1. Inject 1 μL of sample into a gas chromatograph at 210 °C, equipped with hydrogen flame ionization detector [10] (*see* **Note 2**).
2. Run a blank with hexane to check the baseline before and between injections (*see* **Note 5**).

3.3 Fatty Acid Analysis Through Direct Methylation and Gas Chromatography

After integrating the chromatogram peaks using the gas chromatography software, the area of each peak (A) can be obtained to calculate the fatty acid relative amounts and the total fatty acid content.

The peak can be approximated to a triangular shape, considering its height (h) and the base measured by extrapolating the sides of the peak to the baseline (w) [12]. The area is calculated as follows:

$$A_{FA} = \frac{h \times w}{2}$$

The percentage of each fatty acid ($\%FA$) is calculated as follows:

$$\%FA = \frac{A_{FA}}{A_{C16:0} + A_{C16:1t} + A_{C18:0} + A_{C18:1} + A_{C18:2} + A_{C18:3}} \times 100$$

The total FA content (m_{FA}) in the sample per mass of leaf fresh weight (m_{FW}) is calculated as follows, using the internal standard (C17:0) area which corresponds to 20 μg FA:

$$m_{FA} = \frac{\frac{20 \times A_{totalFA}}{A_{C17:0}}}{m_{FW}}$$

To calculate the FA percentages for the APF, the following approach is considered: The leaf fresh weight corresponding to the APF volume used for FA analysis (150 μL) is calculated

Table 1
FA percentages and total FA content in grapevine leaves and APF

	FA percentage						
	C16:0	16:1 *t*	C18:0	C18:1	C18:2	C18:3	m_{FA} (μg/gFW)
Total leaf extract	21.400	2.720	4.379	4.006	21.808	45.686	4119.638
APF	51.299	–	41.653	4.774	1.600	0.674	2170.230

C16:0, palmitic acid; C16:1 t, trans-3-hexadecanoic acid; C18:0, stearic acid; C18:1, oleic acid; C18:2, linoleic acid; C18:3, α-linolenic acid

considering the total APF volume (21 mL) obtained in the extraction from the total amount of leaves (25 g), using three simple rules. FA percentages and total FA content in the total leaf extract and APF are listed in Table 1.

3.4 Apoplast Fluid Purity Assessment Through Fatty Acid Profiling

The activity of cellular enzymes such as malate dehydrogenase or Glc-6-phosphate dehydrogenase are widely used to assess cytosolic contamination of apoplast extracts [8, 13, 14]. The trans-3-hexadecanoic acid (C16:1 *t*) containing phosphatidylglycerol (PG) molecular species are specific to chloroplast membranes [15]. The absence in APF extracts of C16:1 *t* (present in total fatty analysis of leaves) may be used to access apoplast purity (Fig. 1b). In grapevine, a "18:3" plant, C16:1 *t*, is the only FA exclusive to plastid; however, in the case of "16:3" plants such as *Arabidopsis thaliana*, spinach, etc., which contain hexadecatrienoic acid (C16:3) and hexadecadienoic acid (C16:2), these FAs, exclusively present in the plastidial galactolipids monogalactosyldiacylglycerol and digalactosyldiacylglycerol [16], can also be used as intracellular markers (Fig. 1a).

4 Notes

1. It is best to use for APF extractions fully expanded leaves (3–7 leaves from the shoot apex for grapevine) as younger leaves may be more susceptible to mechanical damage that may lead to cell leakage and apoplast contamination.
2. Any gas chromatography equipment equipped with a flame ionization detector (FID) and a column for FAMEs and appropriate software for analysis should be used. In this work a Varian 430-GC gas chromatograph (Varian, Inc., CA, USA) equipped with a hydrogen flame ionization detector set at 300 °C was used. The temperature of the injector was set to 270 °C, with a split ratio of 50. The fused-silica capillary column (50 m × 0.25 mm; WCOT Fused Silica, CP-Sil 88 for FAME; Varian) was maintained at a constant nitrogen flow of

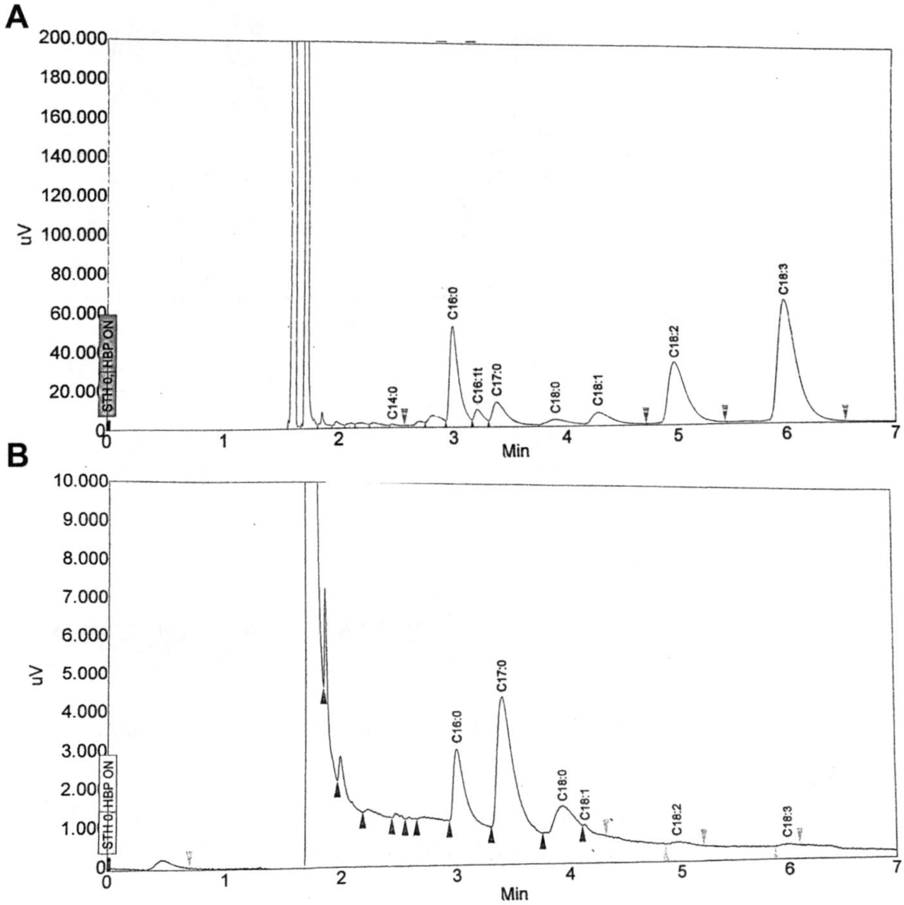

Fig. 1 Gas chromatogram of fatty acid methyl esters from grapevine leaf tissue (**a**) and APF (**b**) FAMEs. C16:0, palmitic acid; C16:1 *t*, trans-3-hexadecanoic acid; C17:0, margaric acid; C18:0, stearic acid; C18:1, oleic acid; C18:2, linoleic acid; C18:3, α-linolenic acid

2.0 mL min^{-1} and the oven temperature set at 190 °C. Fatty acids were identified by comparison of their retention times with standards, and chromatograms analyzed by the peak surface method, using the Galaxy software (Varian, Inc.) as described in [17].

3. The harvested grapevine leaf tissue should be immediately frozen in liquid nitrogen and ground to fine powder in a cooled mortar and a pestle with liquid nitrogen. When turned into fine powder, samples must be weighed. To weigh frozen samples, a 1.5 mL microcentrifuge tube cooled in liquid nitrogen should be used, and the tube should be kept cold during the weighing of the samples. For 50 mg a precision scale should be used.
4. If needed, the FAME samples can be stored for posterior GC analysis. To do so, add to the dry FAMEs 50–100 μL of an

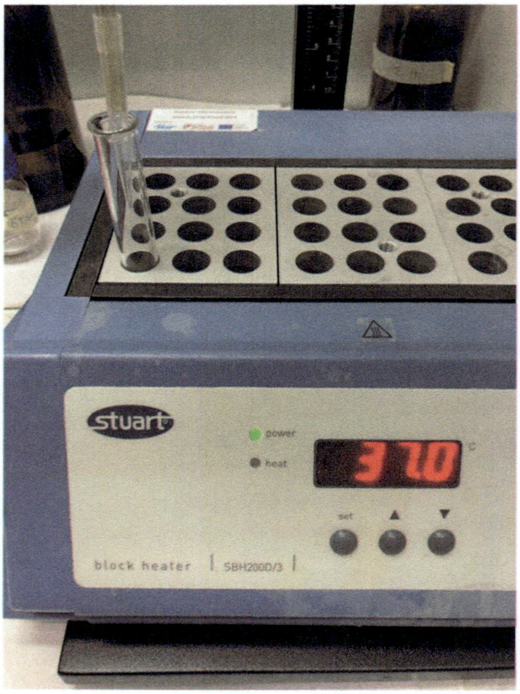

Fig. 2 Samples are dried at 37 °C under nitrogen atmosphere

ethanol-toluene solution (1:4), transfer into a 1 mL vial and dry under nitrogen atmosphere (Fig. 2). Store at −20 °C for up to 1 week.

5. All glassware, including Hamilton syringes, should be thoroughly cleaned, rinsed with acetone, and dried before use.

Acknowledgements

We would like to thank Doctor Joana Figueiredo and Doctor Leonor Guerra-Guimarães for their participation in the development of the grapevine leaf apoplast fluid extraction method.

References

1. Santos RB, Figueiredo A (2021) Two sides of the same story in grapevine-pathogen interactions. J Exp Bot 72:3367–3380
2. Cavaco AR, Matos AR, Figueiredo A (2021) Speaking the language of lipids: the cross-talk between plants and pathogens in defence and disease. Cell Mol Life Sci 78:4399–4415
3. Kachroo A, Kachroo P (2009) Fatty acid-derived signals in plant defense. Annu Rev Phytopathol 47:153–176
4. Laureano G, Cavaco AR, Matos AR, Figueiredo A (2021) Fatty acid desaturases: uncovering their involvement in grapevine defence against downy mildew. Int J Mol Sci 22:5473
5. Cavaco AR, Laureano G, Cunha J et al (2021) Fatty acid modulation and desaturase gene expression are differentially triggered in grapevine incompatible interaction with biotrophs and necrotrophs. Plant Physiol Biochem 163: 230–268

6. Toruño TY, Stergiopoulos I, Coaker G (2016) Plant-pathogen effectors: cellular probes interfering with plant defenses in spatial and temporal manners. Annu Rev Phytopathol 54:419–441
7. Guerra-Guimarães L, Pinheiro C, Chaves I et al (2016) Protein dynamics in the plant extracellular space. Proteomes 4:22
8. Figueiredo J, Cavaco AR, Guerra-Guimarães L et al (2021) An apoplastic fluid extraction method for the characterization of grapevine leaves proteome and metabolome from a single sample. Physiol Plantarum 171:343–357
9. Figueiredo A, Guerra-Guimarães L (2021) Isolation of apoplastic fluid from woody plant leaves: grapevine and coffee as a case study. In: Carrera M, Mateos J (eds) Shotgun proteomics. Methods in molecular biology, vol 2259. Humana, New York, pp 49–57
10. Laureano G, Figueiredo J, Cavaco AR et al (2018) The interplay between membrane lipids and phospholipase A family members in grapevine resistance against *Plasmopara viticola*. Sci Report 28(8):14538
11. Li YH, Beisson F, Pollard M, Ohlrogge J (2006) Oil content of Arabidopsis seeds: the influence of seed anatomy, light and plant-to-plant variation. Phytochemistry 67:904–915
12. Gaul JA (1966) Quantitative calculation of gas chromatographic peaks in pesticide residue analyses. J Assoc Off Anal Chem 49:389–399
13. O'Leary BM, Neale HC, Geilfus CM et al (2016) Early changes in apoplast composition associated with defence and disease in interactions between *Phaseolus vulgaris* and the halo blight pathogen *Pseudomonas syringae* Pv. phaseolicola. Plant Cell Environ 39:2172–2184
14. Rico A, Preston GM (2008) *Pseudomonas syringae* pv. *tomato* DC3000 uses constitutive and apoplast-induced nutrient assimilation pathways to catabolize nutrients that are abundant in the tomato apoplast. Mol Plant-Microbe Interact 21:269–282
15. Paul M (2013) Photosynthesis. Plastid biology, energy conversion and carbon assimilation. *Ann Bot* 111(3)
16. Matos AR, Gigon A, Laffray D et al (2008) Effects of progressive drought stress on the expression of patatin-like lipid acyl hydrolase genes in Arabidopsis leaves. Physiol Plant 134: 110–120
17. Feijão E, de Carvalho R, Duarte IA et al (2020) Fluoxetine arrests growth of the model diatom *Phaeodactylum tricornutum* by increasing oxidative stress and altering energetic and lipid metabolism. Front Microbiol 11:1803

Chapter 15

Identifying Fungal Secondary Metabolites and Their Role in Plant Pathogenesis

Joanna Tannous, Jesse Labbé, and Nancy P. Keller

Abstract

Pathogenic fungi are the main infectious agents of plants. Secondary metabolites produced by these fungi, also recognized as natural products, are key mediators of plant-fungal interactions. Knowledge on the biosynthesis of these metabolites, the accessibility to fungal genome sequences, and the development of gene disruption techniques open up opportunities to identify many more of these metabolites both in vitro and in planta. This methodology chapter gives a detailed systematic approach aiming to discover new natural products from phytopathogenic fungi and characterize their role in triggering plant cell death and plant disease. This approach takes advantage of the global regulation of fungal secondary metabolite production by regulatory proteins reported in various fungal species.

Key words Bioinformatics, Fungal secondary metabolites, Phytopathogenic fungi, Biosynthetic gene clusters, Plant pathogenesis, Fungal-plant interactions

1 Introduction

Fungi possess high-capacity secretory systems and are broadly exploited for the biosynthesis of a vast pool of molecules, ranging from extracellular proteins and polymeric substances to small compounds. These secreted molecules display a broad spectrum of biological properties and can have potential applications in the pharmaceutical, food, and agricultural industries [1]. Secondary metabolites (SMs) (also called natural products) comprise most fungal small compounds and are increasingly recognized for important roles in fungal fitness [2, 3]. Some fungal secondary metabolites were reported to serve as symbiosis agents or communication signals during coexistence with other neighboring microbes or organisms [4–6], while others were described as virulence factors during pathogenic interactions with plants and animals [7]. Different chemical classes of fungal secondary metabolites exist, including terpenes, polyketides, non-ribosomal peptides, and shikimic

Nora A. Foroud and Jonathan A. D. Neilson (eds.), *Plant-Pathogen Interactions*, Methods in Molecular Biology, vol. 2659, https://doi.org/10.1007/978-1-0716-3159-1_15,

acid-derived aromatic compounds [8, 9]. In fungi, the genes encoding secondary metabolite biosynthetic pathways are often spatially clustered at the end of the chromosomes of the producing organism in subtelomeric regions [10]. Those clusters typically comprise genes encoding one or multiple enzymes responsible for synthesizing the core structure of the compound (backbone enzymes), tailoring enzymes that modify the core structure, transporter enzymes, and often regulatory proteins like specific transcription factors (TFs) [9, 11]. Taking advantage of this physical gene clustering in fungi, several computational genomic tools have been established to characterize metabolic gene clusters toward the pursuit of new secondary metabolite discovery [12].

In the past few years, a substantial number of studies have been conducted on phytopathogenic fungi to identify and experimentally characterize new biosynthetic gene clusters and shed light on their role in fungal-plant interactions. These studies have primarily relied on modifying the expression of clustered genes in a native or heterologous setting, by either experimental knockouts or overexpression of clustered genes that may otherwise be silent under standard culture conditions [13]. When coupled with an assessment of changes in metabolite profiles and in planta analyses, these strategies have led to the detection of novel fungal compounds implicated in plant pathogenesis [8]. A remarkable assortment of secondary metabolites used as virulence factors by fungal plant pathogens has been uncovered [7, 14]. These fungal phytotoxins act as mediators of virulence through diverse mechanisms of actions, including but not limited to the undermining of membrane integrity by inhibiting enzymes involved in lipid biosynthesis, induction of apoptosis or programmed cell death, and alteration of gene expression profiles [7]. Important examples are the closely related toxins fumonisin (produced by *Fusarium* spp.) and AAL toxin (produced by *Alternaria alternata*) known to inhibit the enzyme sphinganine N-acetyltransferase (also known as ceramide synthase) which hinder lipid biosynthesis and result in a perturbed membrane ordering and increased membrane permeability [15–17]. Another good example of a fungal SM crucial for virulence on plants is T-toxin, a toxin well characterized for its role in the *Cochliobolus heterostrophus*/maize pathosystem. T-toxin has been shown to be responsible for forming pores in the mitochondrial membrane, which significantly damages the plant cells [18].

This methodology chapter aims to provide a detailed experimental approach (Fig. 1) for secondary metabolite discovery in phytopathogenic fungi by disrupting fungal global regulators associated with global positive or negative regulation of biosynthetic gene clusters (BGCs). Several regulatory proteins have been reported to take part in the regulation of secondary metabolite biosynthesis in fungi, including the pH regulator PacC [19, 20], the carbon catabolite repressor CreA [21, 22], and the nitrogen

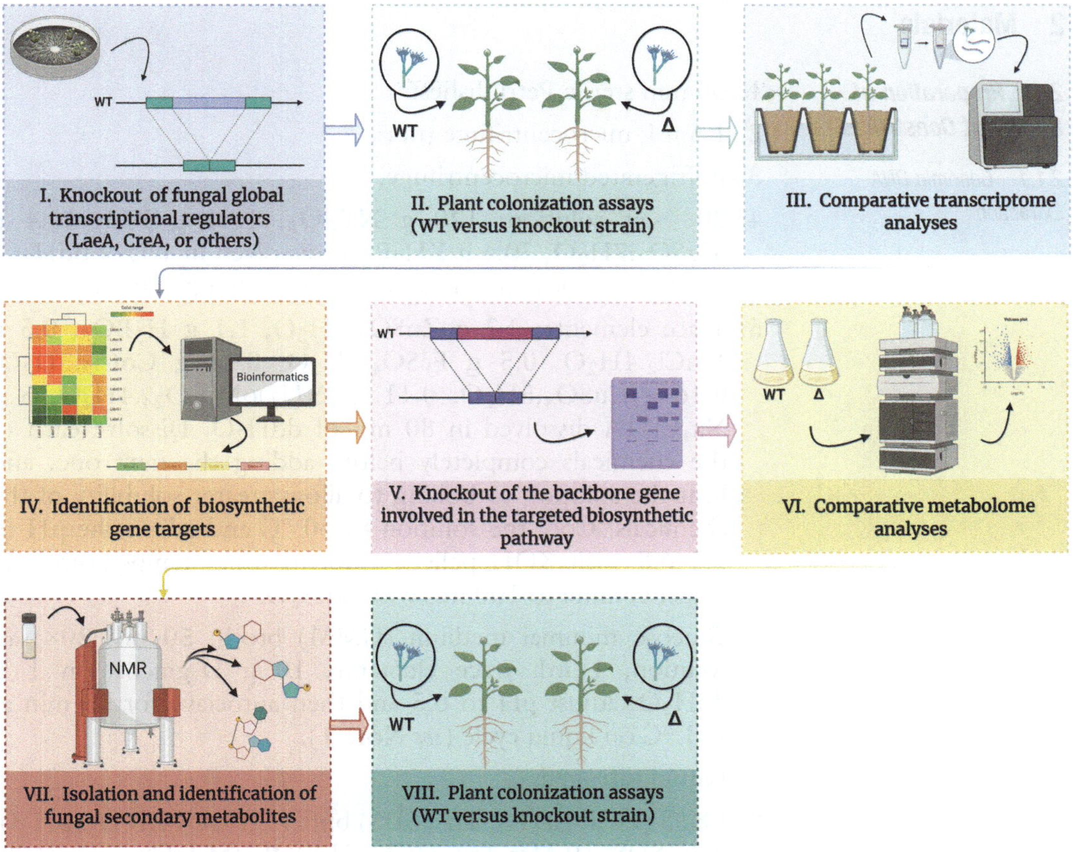

Fig. 1 Schematic summarizing the strategy adopted to identify fungal secondary metabolites and study their role in plant pathogenesis

regulator AreA [23, 24]. However, by far, the most studied transcriptional complex that alters global regulation of secondary metabolites across different fungal genera is the heterotrimeric velvet complex (VelB/VeA/LaeA) [25, 26]. Exploiting the impact of fungal global regulators on fungal pathogenesis and complementing it with metabolome and transcriptome studies will help to narrow down the BGC targets toward identifying new metabolites contributing to the pathogenic behavior of the fungus. Here we describe the bioinformatic and genomic tools employed for the identification of metabolite gene targets, the gene deletion approaches, the analytical methods for metabolite characterization and profiling, and the in vivo approaches for studying secondary metabolites' implication in plant pathogenesis (Fig. 1).

2 Materials

2.1 Preparation of Knockout Constructs

2.1.1 Genomic DNA Extraction

1. 60 mm sterile Petri dishes.
2. 1.5 mL microcentrifuge tubes.
3. Refrigerated microcentrifuge.
4. 20× salt solution: 120 g $NaNO_3$, 10.4 g KCl, 10.4 g $MgSO_4.7H_2O$, 30.4 g KH_2PO_4, dissolved in 1 L of ddH_2O, and then autoclaved and stored at room temperature.
5. Trace elements: 2.2 g $ZnSO_4.7H_2O$, 1.1 g H_3BO_3, 0.5 g $MnCl_2.4H_2O$, 0.5 g $FeSO_4.7H_2O$, 0.16 g $CoCl_2.5H_2O$, 0.16 g $CuSO_4.5H_2O$, 0.11 g $(NH_4)6Mo_7O_{24}.4H_2O$, 5 g Na_4EDTA dissolved in 80 mL of ddH_2O. Dissolve each of the chemicals completely before adding the next one, and heat the solution to boiling to allow greater solubility of the chemicals. Cool the solution to 60 °C and adjust the pH to 6.5–6.8 with KOH pellets. Cool to room temperature and adjust volume to 100 mL with ddH_2O.
6. Glucose minimal medium (GMM) broth: 50 mL 20× salt solution, 1 mL trace elements, 10 g D-glucose in 1 L ddH_2O; adjust pH to 6.5 and then autoclave for 15 min at 121 °C on liquid cycle (*see* **Note 1**).
7. Liquid nitrogen.
8. LETS (LiCl, EDTA, Tris, SDS) buffer: 20 mM EDTA (pH 8), 0.5% SDS, 10 Mm Tris–HCl (pH 8), 0.1 M LiCl.
9. Phenol/chloroform/isoamyalcohol (25:24:1).
10. 70% ethanol (EtOH).
11. 95% EtOH.
12. 10 mg/mL RNase A.

2.1.2 Minipreparation of Plasmid DNA

1. 100 mm sterile Petri dishes.
2. Luria-Bertani (LB) medium: 10 g tryptone, 10 g sodium chloride, 5 g yeast extract. Adjust the pH to 7 with 1 M NaOH solution and the final volume to 1 L with ddH_2O. Add 16 g of agar per 1 L to prepare LB agar plates. Autoclave broth or agar media for 15 min at 121 °C on liquid cycle and store at room temperature. Add appropriate concentration of selected antibiotic for selection once the medium has cooled to 55–66 °C and pour the agar medium into Petri dishes.
3. Resuspension buffer: 50 mM Tris–HCl, pH 8.0, 10 mM EDTA, 20 μg RNase A.
4. Lysis buffer: 200 mM NaOH, 1% SDS.
5. Neutralization buffer: 3 M potassium acetate, pH 5.5.

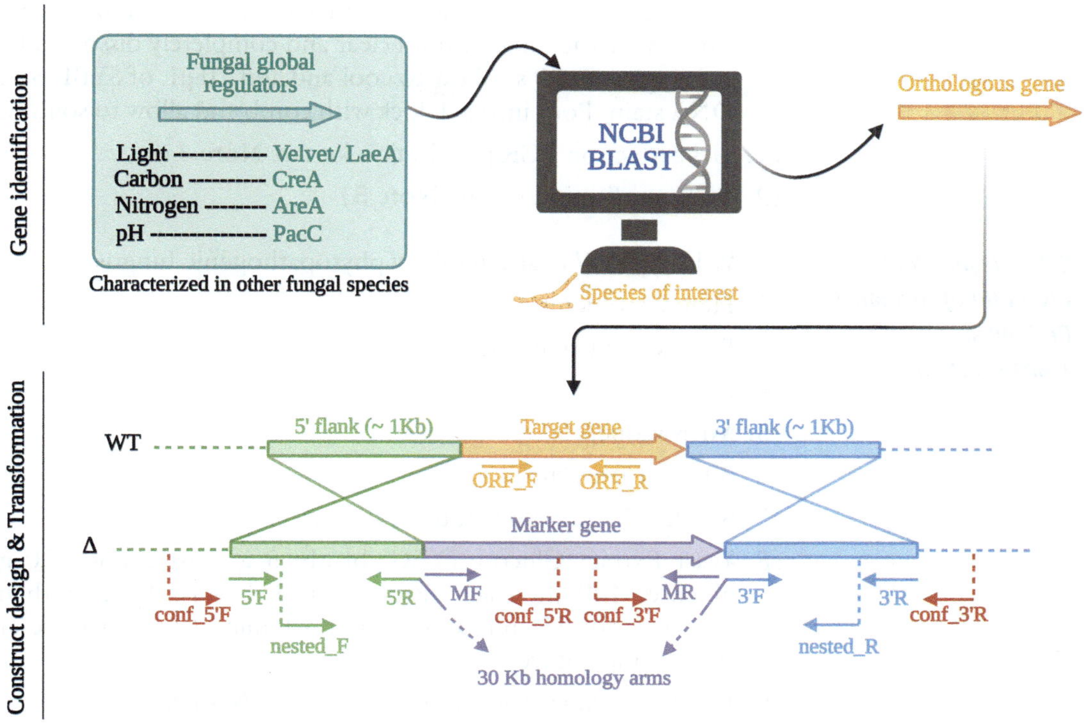

Fig. 2 Schematic of deletion cassette construction using fusion PCR. Primers were designed to amplify flanks both upstream and downstream of the ORF. The positions of the primers used for making the construct and for PCR confirmation of the targeting events are presented by colored arrows. The 5′ forward and 3′ reverse primers contained additional 30 nucleotides to overlap with the marker gene

6. Isopropanol.
7. 70% EtOH.
8. 10 mM Tris–HCl, pH 8.

2.1.3 Knockout Construct Preparation

1. Primer sets designed as shown in Fig. 2.
2. Spectrophotometer.
3. Thermocycler.
4. Gel electrophoresis system.
5. Gel imaging system.
6. dNTP mixture (10 mM of each dATP, dCTP, dTTP, and dGTP).
7. High-fidelity DNA polymerase (*see* **Note 2**).
8. DNA polymerase for long-fragment amplification (*see* **Note 3**).
9. 50X TAE buffer: 50 mM EDTA disodium salt, 2 M Tris base, 1 M acetic acid in 1 L ddH_2O.

10. 0.8% agarose gel: 0.8 g agarose in 100 mL of 1× TAE buffer. Microwave the mixture until clear and completely dissolved by swirling every 30 s. Allow to cool and add 10 μL of SYBR Safe DNA stain. Pour into gel dock with comb and allow to solidify.
11. Gel extraction PCR purification kit (*see* **Note 4**).
12. PCR purification kit (*see* **Note 5**).

2.2 Polyethylene Glycol (PEG)-Mediated Protoplast Transformation

1. Wild-type (WT) strain of any phytopathogenic fungus.
2. Hemocytometer.
3. Refrigerated centrifuge.
4. Miracloth.
5. Microscope.
6. 30 mL Corex tube.
7. Sterile 15 mL round-bottom tubes.
8. Yeast Extract Glucose (YEG) broth: 5 g yeast extract, 20 g glucose, 400 μL trace elements in 1 L ddH_2O, and then autoclave for 15 min at 121 °C on liquid cycle and store at room temperature.
9. Lysing enzymes from *Aspergillus* sp. (*see* **Note 6**).
10. Yatalase (*see* **Note 7**).
11. Osmotic medium buffer: 1.2 M $MgSO_4$, 10 mM Na-PO_4 (pH 5.8) diluted from a 1 M stock of equimolar of mono- and dibasic and filter sterilize it using a 0.2-micron membrane filter.
12. Trapping buffer: 0.6 M sorbitol, 10 mM Tris–HCl, pH of 7.5, and then autoclave for 15 min at 121 °C on liquid cycle.
13. STC buffer: 1.2 M sorbitol, 10 mM Tris–HCl, pH of 7.5, 10 mM $CaCl_2$, and then autoclave for 15 min at 121 °C on liquid cycle.
14. 60% polyethylene glycol (PEG) solution: 60 g PEG 4000 in 100 mL ddH_2O and then autoclave for 15 min at 121 °C on liquid cycle.
15. Sucrose minimal medium (SMM): 50 mL 20× salt solution, 1 mL trace elements, 10 g D-glucose, 218.6 g sorbitol, 15 g agar (for solid media) in 1 L ddH_2O; adjust pH to 6.5 and then autoclave 15 min at 121 °C on liquid cycle (*see* **Note 8**).

2.3 Confirmation of Deletion Mutants

2.3.1 PCR Confirmation

1. Primer sets designed as shown in Fig. 2.
2. Thermocycler.
3. Gel electrophoresis.
4. Gel imaging system.

5. dNTP mixture (10 mM of each dATP, dCTP, dTTP, and dGTP).
6. DNA polymerase.

2.3.2 Southern Blot Confirmation

1. Hybond-XL hybridization membrane.
2. Whatman paper.
3. Saran wrap.
4. Short-wavelength ultraviolet light.
5. Long tweezer.
6. Phosphor-imager.
7. Pyrex glass dish.
8. 5 mL pipette.
9. Hybridization oven.
10. P32 deoxycytidine 5′-triphosphate (dCTP).
11. 5× oligo labelling buffer (OLB): 100 μL 1 M Tris–HCl pH 7.5, 12.5 μL 1 M $MgCl_2$, 250 μL 2 M HEPES buffer pH 6.6, 6.3 μL 10 mM dATP, 6.3 μL 10 mM dGTP, 6.3 μL 10 mM dTTP, 75 μL random primer $p(dN)_6$ (30 A_{260}/mL), bovine serum albumin (BSA) (50 mg/mL) (20 μL) (*see* **Note 9**).
12. 1 M dithiothreitol (DTT).
13. Transfer buffer: 1 M NaCl, 0.4 M NaOH.
14. Pre-hybridization buffer: 280 g SDS, 8 mL 0.5 M EDTA (pH 9.2), 16 mL 85% H_3PO_4, 320 g Na_2HPO_4. Complete the volume to 4 L with ddH_2O.
15. P-60 BioGel:100 mM NaCl, 10 mM Tris–HCl (pH 7), 2.5 mM EDTA. Adjust the pH to 7 with 1 mM HCl.
16. Stop dye: 0.5% Orange G, 0.5% Blue Dextran, 0.5 M EDTA pH 8.
17. 1× SSC, 0.1% SDS: 50 mL 20× SSC, 10 mL 10% SDS solution. Adjust the volume to 1 L with ddH_2O.
18. 20x SSC: 175.3 g NaCl, 88.2 g $Na_3C_6H_5O_7 \cdot 2H_2O$. Adjust the pH to 7 with 1 mM HCl and adjust the volume to 1 L with ddH_2O.

2.4 Host Colonization Assays

2.4.1 Preparation of Fresh Spore Suspension

1. Fresh culture on potato dextrose agar (PDA) of the WT strain of the fungal species of interest.
2. 0.01% Tween-20.
3. Sterile L-shaped cell spreaders.
4. Hemocytometer.

2.4.2 Colonization of Plant Species or Fruit and Vegetable Commodities

1. Plant or vegetable materials.
2. 0.2% chlorine water.
3. Deepot cells or pots.
4. Support trays.
5. Soil.
6. Sterile toothpicks.
7. Sterile spray bottles.

2.5 RNA Extraction and Sequencing Experiments

1. Plant RNA isolation kit (*see* **Note 10**).
2. Liquid nitrogen.
3. Mortar and pestle.
4. 2 mL microcentrifuge tubes.
5. 50 mL tubes.
6. Ethanol 100% (200 proof).
7. Amplification grade DNase I, RNase free.
8. Diethylpyrocarbonate (DEPC)-treated water.

2.6 Comparative Metabolome Analyses

1. Handheld homogenizer.
2. HPLC vials.
3. UHPLC-MS system.
4. PDA medium.
5. Yeast extract sucrose (YES) agar medium: 30 g glucose, 5 g yeast extract, 15 g agar in 1 L ddH_2O, and autoclave for 15 min at 121 °C on liquid cycle.
6. Czapek concentrate: 30 g $NaNO_3$, 5 g KCl, 5 g $MgSO_4$. $7H_2O$, 0.1 g $FeSO_4$. $7H_2O$, 0.1 g $ZnSO_4$. $7H_2O$, 0.05 g $CuSO_4$. $5H_2O$ in 1 L ddH_2O.
7. Czapek yeast agar (CYA) medium: 1 g K_2HP0_4, 10 mL Czapek concentrate, 5 g yeast extract, 30 g sucrose, 15 g agar in 1 L ddH_2O, and autoclave for 15 min at 121 °C on liquid cycle.
8. Ethyl acetate.
9. Acetonitrile/water (80:20 *v*/*v*).

3 Methods

3.1 Knockout of Fungal Global Transcriptional Regulators

3.1.1 Design of Knockout Constructs

1. Select the gene of interest among the list of fungal global regulators reported to play an important role in secondary metabolite biosynthesis (*see* **Note 11**).
2. Identify the putative ortholog of the target gene in the strain of interest by conducting a BLAST search against sequences in the

GenBank database using as a template a characterized sequence of this gene in other fungal species.

3. Check the presence of conserved domains in the putative protein sequence to affirm family membership.
4. Design the knockout construct by selecting two 1-kb segments of gDNA called "homology arms" that correspond to the 5′ and 3′ flanking ends of the gene. Those two segments should flank the selectable gene marker (Fig. 2).
5. Design two primer sets (5′F/R and 3′F/R) to amplify the 1-kb DNA regions flanking the gene of interest and a primer set (M-F/R) to amplify the selectable marker gene. The reverse primer (5′R) designed to anneal to the 5′-flanking sequence and the forward primer (3′F) designed to anneal to the 3′--flanking sequence should include about 30-kb homology arms corresponding, respectively, to the start and the end of the selection marker cassette (Fig. 2).
6. Design a nested primer pair (Nested_F/R) to amplify the whole deletion cassette during the third-round PCR.

3.1.2 Genomic DNA Extraction

1. Pour 15 mL of GMM broth supplemented with yeast extract (at 5 g/L) into a sterile 60 mm Petri dish.
2. Inoculate the medium with a loopful of spores from the WT strain.
3. Incubate the culture in the dark at the appropriate temperature until a sufficient biomass was formed (*see* **Note 12**).
4. After incubation, pour off the medium, collect the hyphal mat using sterile toothpicks, and place it between paper towels to squeeze out the remaining liquid.
5. Collect the mycelium in a 1.5 mL microcentrifuge tube and freeze-dry mycelium overnight (*see* **Note 13**).
6. Break up the freeze-dried mycelium into a fine powder with a microcentrifuge pestle.
7. Add 700 μL of LETS buffer to the ground powder and vortex to wet the sample. Leave the sample on bench for 5 min.
8. Add 700 μL of phenol/chloroform/isoamyl alcohol (25:24:1). Mix thoroughly by vortexing and by inverting the tube 4–6 times. Allow the sample to set on the bench for 5 min and then centrifuge at 4 °C and 13,000 rpm for 10 min. Repeat this step twice to remove protein and other cellular contaminants.
9. Transfer the supernatant to a new tube and add 1 mL of 95% EtOH. Mix gently by inverting the tubes 4–6 times and centrifuge at 4 °C and 13,000 rpm for 10 min to pellet genomic DNA (gDNA) (*see* **Note 14**).

10. Discard the supernatant carefully without disturbing the pellet and add 1 mL of cold 70% (*v/v*) EtOH to the pellet. The pellet should be completely dispersed in EtOH.
11. Centrifuge the sample for 5 min to pellet gDNA at room temperature (RT).
12. Discard supernatant (be careful not to lose the pellet) and dry pellet at RT for 5 min (*see* **Note 15**).
13. Resuspend the pellet in 40 μL of sterile ddH_2O and add 2 μL RNase A (10 mg/mL).
14. Incubate the sample for 30 min at 65 °C to heat inactivate DNase and digest RNA.
15. Store the gDNA sample at −20 °C for later use.

3.1.3 Minipreparation of Plasmid DNA

1. If the selectable marker to be used for construction of gene deletion cassette is contained within a plasmid, grow the bacterial culture from the glycerol stock to isolate individual colonies on LB agar plates supplemented with the appropriate antibiotic for selection.
2. Incubate the plate with freshly plated bacteria overnight at 37 °C.
3. Pick a single colony to inoculate 3 mL of LB broth supplemented with the same selection antibiotic.
4. Grow the culture at 37 °C and 200 rpm overnight.
5. Harvest the bacterial cells by centrifugation at 12,000 rpm for 30 s and discard the supernatant without disturbing the pellet.
6. Resuspend pelleted bacterial cells in 100 μL of resuspension buffer and transfer to a 1.5 mL microcentrifuge tube. The bacteria should be resuspended completely by vortexing until no cell clumps remain.
7. Add 100 μL of neutralization buffer and mix thoroughly by inverting the tube four to six times. Do not vortex at this step as this may result in shearing of the genomic DNA.
8. Add 125 μL of neutralization buffer and mix immediately by inverting the tube four to six times to avoid precipitation. At this step, the solution should become cloudy.
9. Centrifuge for 10 min at 13,000 rpm at 4 °C; a white pellet will form.
10. Collect the supernatant and transfer it to a new 1.5 mL microcentrifuge tube.
11. Add 200 μL of isopropanol and mix well to pellet plasmid DNA. Leave the sample for a couple of minutes on ice.
12. Collect DNA pellet by centrifuge at 13,000 rpm, 4 °C, for 2 min and decant supernatant.

13. Add 500 μL of 70% ethanol to wash the pellet.
14. Centrifuge again for 30 s at 13,000 rpm, 4 °C, to collect the DNA pellet and remove the supernatant.
15. Air-dry the pellet for 3 min on the bench and resuspend it in 50 μL of 10 mM Tris–HCl buffer, pH 8.0. Sterile ddH_2O can be used instead for resuspension of the DNA pellet.

3.1.4 PCR-Mediated Generation of Gene Knockout Construct

1. Three rounds of PCR will be required. For the first round, using WT strain gDNA amplifies the upstream flanking region (5′ 1-kb region) using the primer pair 5'F/R and the downstream flanking region (3′ 1-kb region) using the primer pair 3' F/R. Using the adequate plasmid DNA amplifies the marker gene (e.g., *hph, argB, pyrG*, or others) using the primer pair M F/R. The PCR conditions for the first-round PCRs are described in Table 1.
2. Check the PCR products on a 0.8% agarose gel.
3. If the gel electrophoresis shows a single band of the expected size, purify the PCR product using a PCR purification kit following manufacturer's instructions (*see* **Note 16**).
4. If multiple bands are amplified for the same sample, optimize the cycling conditions as follows: (i) increase the annealing temperature as a strategy to reduce nonspecific primer binding (gradient PCR can be used for this) and (ii), if the nonspecific products are shorter than the target, increase the annealing time, or reduce it if they are longer.
5. If the nonspecific bands are still showing, cut the band of interest and proceed with gel extraction using a gel extraction kit following the manufacturer's instructions (*see* **Note 17**).
6. Estimate the concentration of the purified PCR products and conduct the second-round PCR using the conditions described in Table 2.

Table 1
Thermocycling conditions for the first-round PCR

PCR setup		Thermocycling conditions			
Template DNA	10–50 ng	Step	Temperature	Time	Cycles
Forward primer (20 pmol)	1 μL	Initial denaturation	95 °C	3:00	1
Reverse primer (20 pmol)	1 μL	Denaturation	95 °C	0:30	34
dNTP (10 mM)	1 μL	Annealing	Tm °C[a]	0:30	
10X Pfu II buffer	5 μL	Elongation	68 °C	1:00/kb	
Taq polymerase (Pfu II)	1 μL	Final elongation	68 °C	1:00/kb +2:00	1
ddH_2O	Up to 50 μL				

[a]The temperature of this step depends on the melting temperature of the primer-template hybrid

Table 2
Thermocycling conditions for the second-round PCR

PCR setup		Thermocycling conditions			
5′ Flank (1CN[a])	x μL (100–150 ng)	Step	Temperature	Time	Cycles
Selection marker (2CN)	y μL	Initial denaturation	95 °C	3:00	1
3′ Flank (1CN)	z μL	Denaturation	95 °C	0:30	12
dNTP (10 mM)	1 μL	Annealing	Tm °C	2:00	
10X Pfu II buffer	5 μL	Elongation	68 °C	1:00/kb	
Taq polymerase (Pfu II)	1 μL	Final elongation	68 °C	1:00/kb +2:00	1
ddH_2O	Up to 50 μL				

[a]*CN* copy number

Table 3
Thermocycling conditions for the third-round PCR

PCR setup		Thermocycling conditions			
Second round PCR product	5 μL	Step	Temperature	Time	Cycles
Nested primer F (20 pmol)	1 μL	Initial denaturation	95 °C	3:00	1
Nested primer R (20 pmol)	1 μL	Denaturation	95 °C	0:30	34
dNTP (10 mM)	1 μL	Annealing	Tm °C	2:00	
10X Roche buffer	5 μL	Elongation	68 °C	1:00/kb	
Taq polymerase (Roche)	1 μL	Final elongation	68 °C	1:00/kb +2:00	1
ddH_2O	up to 50 μL				

7. Perform the third-round PCR following the thermocycling conditions described in Table 3, using as a template the product from the second-round PCR and the nested primer set.
8. Purify the PCR product as reported in **step 2** and estimate the concentration using a spectrophotometer.
9. Before using the PCR product for transformation, check the full size of the construct with at least two restriction enzyme digestions.

3.1.5 PEG-Mediated Protoplast Transformation

1. Grow the WT fungus on appropriate medium (e.g., PDA) for at least 5 days at 25 °C. Collect fresh spores by scraping using 10 mL of 0.1% Tween-80 in ddH_2O. Determine spore concentration using a hemocytometer.
2. Inoculate 500 mL of YEG broth with about 10^9 fresh spores and shake at 25 °C and 200 rpm for approximately 10 h. A proper culture should have an abundance of young germlings in small aggregates.

3. Harvest the germlings by filtration through a sterile Miracloth and wash the mycelia with about 50 mL mycelium wash solution.
4. Transfer germlings directly into a 125 mL Erlenmeyer containing 30 mg lysing enzymes from *Aspergillus* sp. and 20 mg Yatalase in 10 mL of osmotic medium.
5. Shake the mixture at 80 rpm and 28 °C for 4 h. Using a microscope, evaluate the progress on cell wall digestion and observe the resulting protoplast isolation. If protoplasts are forming rapidly, 2–3 h is sufficient (*see* **Note 18**).
6. Pour cells directly in a sterile 30 mL Corex tube and overlay them gently with 10 mL of trapping buffer. Centrifuge at 5000 rpm for 15 min at 4 °C.
7. Collect the protoplasts gently from the interface with a pipette and transfer them to a sterile 15 mL tube.
8. Dilute the protoplast suspension with an equal 1 volume of ice-cold STC buffer. Centrifuge at 3000 rpm for 10 min.
9. Decant the supernatant gently without disturbing the protoplast pellet and resuspend it in 1 mL of STC buffer. Estimate the number of protoplasts using a hemocytometer.
10. Centrifuge the protoplast suspension at 3000 rpm for 5 min and discard the supernatant.
11. Resuspend the protoplast pellet in STC buffer at a concentration of 10^8 to 10^9 and store on ice.
12. Dilute the DNA construct (prepared in Subheading 3.1) to 3–5 μg in STC buffer while keeping the final volume at 100 μL in a sterile 15 mL round-bottom tube.
13. Add 100 μL of protoplasts to the DNA and incubate the mixture for 50 min on ice (*see* **Note 19**).
14. Remove the tubes from ice and add 1.25 mL of PEG solution to each tube, mix gently, and incubate at RT for 20 min.
15. Add 5 mL of STC buffer and mix gently.
16. Plate on selective media (at least eight plates per transformation). Take about 500 μL to 1 mL per plate. Incubate right side up overnight to let dry.
17. Check the plates periodically for the appearance of transformants.
18. Randomly select several transformants and transfer them with sterile toothpicks on new SMM selection plates to undergo a second screening round.
19. Reculture the isolated transformants on GMM broth and extract gDNA following the same technique described previously in Subheading 3.1.2.

3.1.6 Identification of Positive Transformants by PCR

1. Conduct a PCR amplification following the same thermocycling conditions described previously in Subheading 3.1.4, **step 1**, using the primer set ORF-F/R designed to amplify the open reading frame (ORF) of the knocked-out gene. Use as a positive control gDNA from the WT strain. Positive transformants should not give any band (*see* **Note 20**).
2. It is possible to obtain false positives; therefore, it is recommended to follow up with two PCR confirmations using the primer sets conf_5'F/R and conf_3'F/R. One of the primers in each set is designed outside the knockout construct and the other on the marker gene. The position of all the primers used for screening of transformants is shown in Fig. 2.
3. Run all PCR products on a 1% agarose gel.
4. Identify the PCR-positive transformants and follow up with Southern blot analysis to confirm the single insertion of the deletion cassette into the WT genome.

3.1.7 Southern Blot Confirmation Screening

1. Digest 5 μg of gDNA from transformants and the positive control (WT strain) using specific restriction enzymes that surround the mutant gene locus and will allow to distinguish genetic variability between the mutant and the WT. The intended restriction is a critical step in Southern blot analysis. A recommended Southern blot design is shown in Fig. 3 (*see* **Note 21**).
2. Run digested gDNA on a 1% agarose gel at 5–7 volts per cm.
3. Verify the quality of the digestion on a gel imaging system. It should appear as a continuous smear of DNA, rather than distinct bands.
4. Rinse the gel in ddH_2O and then incubate it in 0.25 M HCl for 10 min (with gentle shaking) until the lower band turns yellow. Remove HCl and rinse with ddH_2O.
5. Incubate the gel for 10 min (with gentle shaking) in transfer buffer. The lower bands will turn blue again.
6. Prepare blotting assembly as diagrammed in Fig. 4 and fill the pyrex dish with transfer buffer. Remove air bubbles between layers by gently rolling a 5 mL pipette across the top layer before adding the next layer of the assembly.
7. Allow the capillary transfer to sit overnight, for at least 12 h (*see* **Note 22**).
8. Disassemble the blot the following day. Before removing the membrane from the gel, note the position of the wells on the membrane with a pencil.
9. Dry the Hybond-XL hybridization membrane completely in an oven at 80 °C for 10 min, and then cross-link the blotted DNA

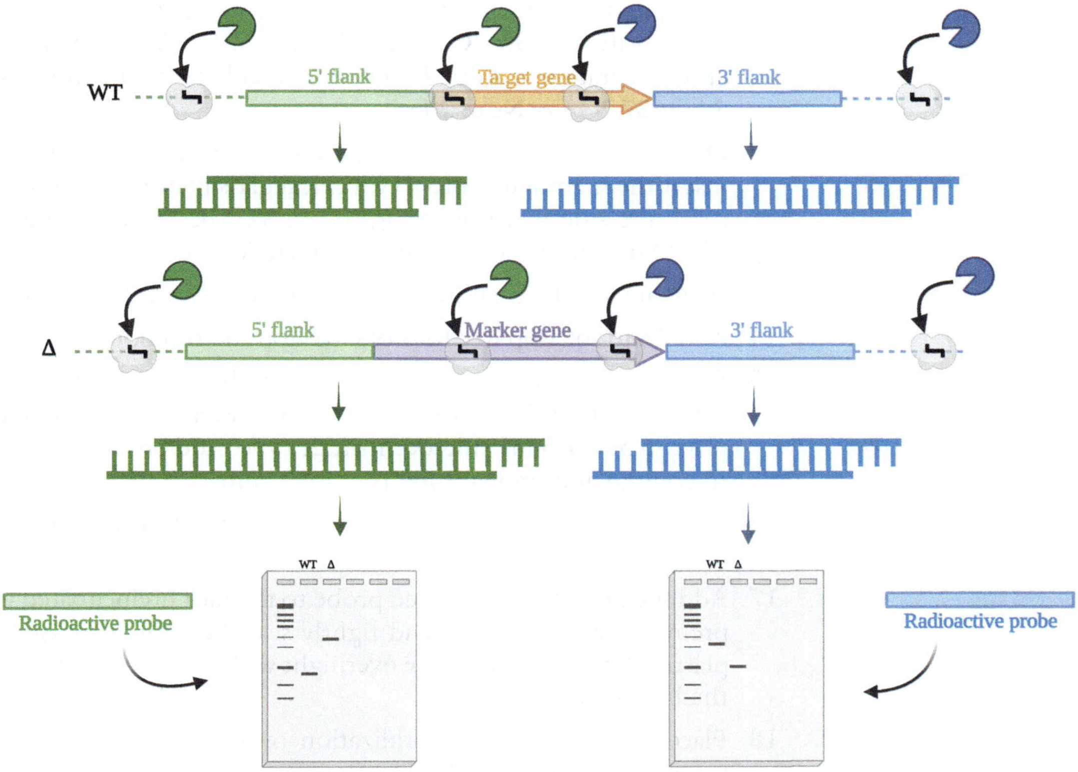

Fig. 3 Schematic diagram of the Southern blot design recommended to screen for homologous targeting events

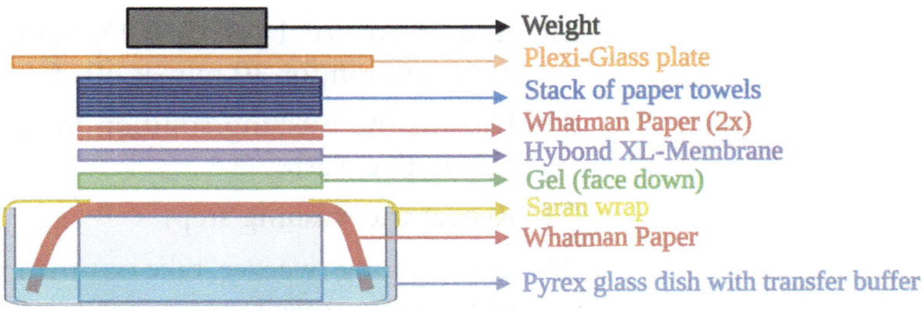

Fig. 4 Schematic illustration of the Southern blot transfer assembly (*see* **Note 23**)

to the membrane by exposure to short-wavelength ultraviolet light.

10. Wrap the blot in saran wrap and store it at −20 °C until time of use.
11. For making southern blot probes, use the 5′ and 3′ flank-DNA fragments synthesized earlier in Subheading 3.1.4, **step 1**. Suspend 300 ng of each of the fragments, separately, in 17 μL ddH_2O. Boil the DNA for 5 min, and then place it on ice for

20 seconds. To radioactively label the probe, add 5 μL of 5x OLB, 3 μL of P32 dCTP, and 0.5 μL of DNA polymerase I (Klenow fragment from *E. coli*), and incubate the mixture for 3 h at 37 °C (*see* **Note 24**).

12. One hour before the end of the incubation period, carefully insert the membrane into a cylindrical glass hybridization tube (with the side that was facing the gel on the inside) and add 35–30 mL of pre-warmed prehybridization buffer.
13. Incubate at 60 °C within hybridization over for 30–45 min.
14. Add 25 μL of stop dye to probe to stop the reaction.
15. Prepare the probe cleanup G-50 column by adding 800 μL of P-60 BioGel. Spin 3000 rpm for 1 min. Remove column and place in new 1.5 mL microcentrifuge tube. Add probe contents to column and spin at 3000 rpm for 1 min.
16. Boil flow-through for 5 min; be sure to cap the tube. Incubate on ice for 1–2 min.
17. Add the radioactively labeled probe to the blot preincubated in prehybridization buffer, and tightly seal the container with a plastic O-ring cap. Incubate overnight with rotation at 60 °C in the hybridization oven.
18. Place the tube in the hybridization oven, and incubate with rotation overnight at 60 °C.
19. The following day pour off the hybridization solution into a labeled falcon tube and store it at −20 °C (hybridization solution can be reused for a few weeks).
20. Add 20 mL of 1× SSC, 0.1% SDS, to the tube and incubate with rotation for 30 min at 60 °C.
21. Discard the washing solution in an appropriate container for radioactive waste.
22. Repeat the washing step.
23. Using a long tweezer, take out the membrane gently and dry it using paper towels.
24. Wrap the dry membrane with saran wrap and place it in a cassette that contains a phosphor screen. Make sure that the side of the membrane that has the transferred DNA is facing the screen.
25. Close the cassette to allow the radiation from the membrane to produce an image on the screen and let it set overnight.
26. Take the screen from the cassette and place on the phosphor imager and analyze readout with scanning software.

3.2 In Vivo Colonization Assays

3.2.1 Preparation of Fresh Spore Suspension

1. Grow the WT and mutant strains on a rich medium (e.g., PDA) for at least 5 days at the appropriate temperature.
2. Harvest spores from each strain by flooding the cultures with 5 mL of sterile water containing 0.01% (v/v) of Tween-20 and dislodging spores from the mycelia using a sterile plastic spreader.
3. Count spores using a hemocytometer and prepare spore suspensions at a concentration of 10^6 spores/mL.

3.2.2 Colonization of Plant Species

In planta experiments can be conducted through above- or belowground inoculations of fungal pathogen(s). Koch's postulates are used to ensure that the fungus causing the symptoms and recovered from the infected plant tissue is the same one used for inoculations. Aboveground applications are typically used for foliar fungal pathogens or when investigating late-stage disease development, whereas belowground applications are used for soilborne pathogens or when studying seedling disease or early-stage disease development. The decision to use either one of those applications depends on the pathogens' infection route, the plant development stage, and the experimental question the researcher is trying to answer.

1. Surface sterilize Deepot cells and racks, pots, or cell trays with chlorine water (20%) for 10 min, and then wash twice with water (*see* **Note 25**).
2. Choose a specific type of soil (clay, sandy, silty, peaty, chalky, or loamy) to simulate the conditions where the intended plants were originally found and the disease was observed.
3. Layer 2–4 inches of soil evenly on an autoclaving tray and autoclave it at 121 °C for 30 min at 15 psi. Add 25 mL of water to the soil mixture before autoclaving. Optionally, mix the autoclaved soil and repeat the autoclaving process. Pasteurized soil is also an option for retaining some of the soil microbiome. Pasteurization conditions are 82 °C for 30 min.
4. After autoclaving, add primary nutrients (macronutrients) to the soil, which include carbon, hydrogen, nitrogen, phosphorus, and potassium. Micronutrients such as calcium, magnesium, and sulfur can also be added. Hoagland solution is commonly used in this kind of experiment (*see* **Note 26**).
5. Before inoculation, allow the plants to grow and reach a stage prior to the optimal stage for disease observation (vegetative, budding, flowering, ripening).
6. For aboveground inoculations, prepare a fresh spore suspension of the fungus of interest as previously stated in Subheading 3.2 and transfer it under sterile conditions into a sterilized spray bottle.

7. Once the plants have reached the desired growth stage, spray the leaves and stems of the plant until the surface of the leaf or stem is covered with solution. Avoid over-spraying the plants and dripping. For some plant pathogen associated with stem rots, soaking sterile toothpicks in inoculum and gently pricking the stem of the host plant are necessary [27].
8. Set the growth chamber or greenhouse to the desired environmental settings conducive for disease progression and observations (controlled temperature, humidity, lighting, etc.), and incubate the plants for a specific time period.
9. Measure plant growth parameters (e.g., stem width and height, leaf width and length, chlorophyll content, height, and diameter of the plant) and disease severity using published scales (i.e., Horsfall-Barratt scale) or tools that are commercially available. It is recommended to determine total dry plant biomass weight (stem and leaves) when comparing the WT to the mutant.
10. To complete Koch's postulates, surface-sterilize the plant material and recover fungal pathogen(s) from the infected plant tissue.
11. Determine the identity of the pathogen through morphological and molecular sequencing techniques (amplification of specific molecular markers).
12. For belowground inoculations, prepare the plant containers, soil, and spore suspension following the same **steps 1–5**.
13. Add 1 mL of spore suspension to 1 g of autoclaved soil used (*see* **Note 27**).
14. Depending on the seed type, seeds are planted at about 1–3 inches deep into the Deepot cell or pot and 0.5–1 inch deep into the cell trays. Alternatively, pre-germinated seeds in sterile conditions can be used.
15. Incubate the samples at the optimal growth- and disease-inducing conditions.
16. Measure root parameters using appropriate software (e.g., WinRHIZO https://regent.qc.ca/assets/winrhizo_software.html) and record disease development on roots and plant growth parameter disease severity using published scales (i.e., Horsfall-Barratt scale) or tools that are commercially available. It is recommended to determine the total dry root weight when comparing the WT to the mutant.
17. Complete Koch's postulates as described for aboveground applications.

3.2.3 Colonization of Fruit or Vegetable Commodities

1. Sterilize the surface of the fruits or vegetables by soaking them in chlorine water (0.2%) for 2 min. Perform at least six technical replicates per strain.

2. Rinse the samples for 2 min using clean running water to remove residual chlorine.
3. Create one artificial wound per sample of approximately 0.5 cm depth using a sterile toothpick.
4. Prior to inoculation, let the wounded samples set for 30 min under sterile conditions.
5. Inject 10 μL of the fresh spore suspension (10^6 spores/mL) of either the WT or the knockout mutant strains into the wounded sites.
6. Incubate the inoculated samples at the appropriate temperature for 2 weeks and record the diameters of the rotten spots daily.
7. If the mutant strain showed a significantly reduced or increased virulence on plants (or vegetables) compared to the WT strain, proceed to the next step which consists of looking at transcriptomic and metabolomic data to identify putative secondary metabolites responsible for changes in pathogenesis. If no significant differences were observed between the WT and the mutant data, pick another candidate, from the list of regulatory proteins reported to have an impact on fungal secondary metabolism and pathogenicity, to conduct this study.

3.3 Comparative Transcriptome Analysis: RNA Extraction from Infected Plant or Vegetable Materials

1. Harvest tissues from plants, fruits, or vegetables infected by either the WT or mutant strains and submerge samples in liquid nitrogen to prevent RNA degradation.
2. Grind the samples to a fine powder in liquid nitrogen using a mortar and pestle and transfer the powder into sterile 50 mL tubes pre-chilled on ice.
3. After liquid nitrogen evaporation, quickly weigh the recommended amount of tissue powder in 2 mL microcentrifuge tubes pre-chilled on ice or liquid nitrogen.
4. Use plant RNA isolation kit according to manufacturer's instructions.
5. Store purified RNA at −20 °C for short-term use or − 80 °C for long-term use.
6. Determine the concentration and quality of the RNA samples using a spectrophotometer.
7. Prepare and submit the isolated RNA samples for sequencing following the submission requirements. Several papers have been published in recent years describing in detail the guidelines to follow for RNA sequencing from sample preparation to library construction to data analysis [28–30].
8. Randomly select some genes for follow-up and validation of their expression using qPCR.

3.4 Identification of Biosynthetic Gene Targets for Knockout Assays

1. If a whole-genome sequence is available for the fungal strain used in the study, run the genome sequence on available webtools, e.g., antiSMASH (Antibiotics and Secondary Metabolite Analysis Shell), which is the most commonly used pipeline for identification of biosynthetic loci responsible for production of various classes of secondary metabolite compounds [9, 31]. If a whole-genome sequence for the strain of interest is not available, it is difficult and impossible in this case to ID biosynthetic gene clusters (BGCs) before whole-genome sequencing. In recent years, this technology has become much easier, faster, and less expensive. Recently, Jung et al. have provided a detailed stepwise procedure that can be followed for genome sequencing, assembly, and annotation [32].
2. Use the list of significantly downregulated or upregulated genes to identify specific BGCs that might be related to the observed shifts in secondary metabolite production.
3. Search within the antiSMASH-predicted clusters for biosynthetic genes determined to be differentially regulated based on RNA-seq data. Finding overlapping biosynthetic genes between antiSMASH and RNA-seq analyses will help identify putative targeted BGC(s) that could be playing key roles in fungal pathogenesis (Fig. 5).
4. Perform a targeted deletion of the backbone biosynthesis gene of the identified BGC(s) following the same approach described earlier in Subheading 3.1.

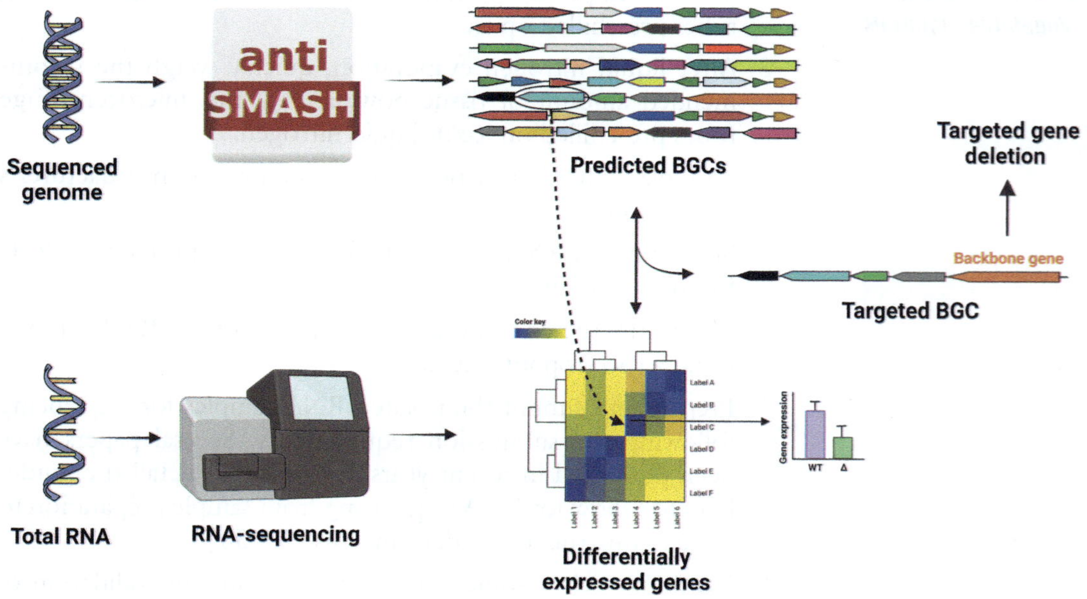

Fig. 5 Schematic summarizing the strategy adopted for identification of biosynthetic gene targets for knockout assays

3.5 Comparative Metabolome Analyses

1. Culture the biosynthetic mutant and the WT strain by centrally inoculating 10^6 fresh spores on different agar media known to promote production of secondary metabolites in the fungus (e.g., PDA, CYA, YES agar, etc.). Prepare at least five technical replicates per strain and medium.
2. Incubate cultures at the suitable temperature for up to two weeks.
3. After incubation, freeze-dry the agar cultures and grind them into 5 mL of sterile water using a handheld homogenizer.
4. Add 5 mL of ethyl acetate to the aqueous mixture and mix them thoroughly by inverting the tubes several times to form a transiently homogeneous suspension. Repeat this step three to four times.
5. Centrifuge the samples at 3500 *g* for 10 min to ensure a complete separation of the organic and aqueous phases.
6. Transfer the organic layers into new glass vials and then dry them in a speed vacuum at room temperature.
7. Dissolve the crude extracts in acetonitrile/water (80:20 *v/v*) at a concentration of 1 mg/mL.
8. Filter the samples through syringe filters with nylon membrane (0.45 μm) into 1 mL HPLC vials for UHPLC-MS analysis.
9. Verify that the mutant strain was unable to produce the targeted compound by comparing its chromatogram to that of the WT strain.

3.6 Isolation and Identification of Fungal Secondary Metabolites

1. If the targeted metabolite(s) has not been identified or characterized before, grow the WT fungus at a large scale (20 Petri dishes, 150 mm diameter, containing the appropriate medium used earlier) to conduct NMR analysis.
2. Allow the cultures to grow for 20 days at the appropriate temperature.
3. At the end of the incubation period, combine agar media from the 20 plates, freeze-dry them, and extract metabolites following the same approach described in Subheading 3.5 (*see* **Note 28**).
4. To isolate and purify the targeted metabolite, run the preparative HPLC chromatography following the same conditions used for the analytical HPLC chromatography. It is possible that it could take several rounds of preparative fractionation to obtain the necessary concentration of the metabolite in question for NMR. Refer to the following papers for more detailed description of the isolation and characterization of new metabolites by NMR [33, 34].

3.7 In Vivo Colonization Assays

1. After confirming in vitro, the loss of production of the targeted metabolite in the mutant, repeat the in vivo colonization assays as described in Subheading 3.2 using the biosynthetic mutant versus the WT strain.
2. Record the disease progress on plants/vegetable material over time.
3. Apply an appropriate statistical test for comparison between the two groups.

4 Notes

1. For DNA extraction, supplement the GMM broth with yeast extract (5 g/L).
2. For the first- and second-round PCRs, we recommend the use of the PfuUltra II Fusion HS DNA polymerase (Agilent Technologies, Santa Clara, CA, USA; cat. no. 6000672).
3. For the third-round PCR, we recommend the use of the Expand Long Template Polymerase (Roche Applied Science, Penzberg, Germany; cat. no. 11759060001).
4. We recommend the use of the QIAquick Gel extraction kit (Qiagen, Valencia, CA, USA; cat. no. 28706).
5. We recommend the use of the QIAquick PCR purification kit (Qiagen, Valencia, CA, USA; cat. no. 28104).
6. We recommend the use of the lysing enzymes from *Aspergillus* sp. (Sigma Aldrich, St. Louis, MS, USA; cat. no. L3768).
7. We recommend the use of Yatalase (Takara Bio, Shiga, Japan; cat. no. T017).
8. For selection of positive transformants, add or subtract the appropriate supplements (antibiotics, amino acids, or others) from the SMM to support or select against the growth of particular genotypes.
9. To prepare the 5× OLB solution, remove 1 mL aliquots of each component and heat them individually at 65 °C for 10 min. After combining all the components, add 25 μL of 1 M DTT and store the mixture in small aliquots at −20 °C.
10. We recommend the use of the Spectrum Plant Total RNA Kit (Sigma Aldrich, St. Louis, MS, USA; cat. no. STRN50).
11. Several fungal global regulators have been identified and reported having an impact on fungal secondary metabolism. These global regulatory proteins can be used in studies. Among them are the two key members of the velvet protein complex, VeA, and the methyl transferase LaeA, the carbon catabolite repressor CreA, and the epigenetic reader SntB.

12. Depending on the fungal species, the optimal incubation temperature can vary between 25 °C and 37 °C.
13. Poke holes in the lid of the tube (with a syringe needle or a scalpel) and plunge the sample promptly in liquid nitrogen before lyophilization.
14. To precipitate DNA, mix the sample gently by inverting the tube several times. It is key at this step not to vortex or mix the sample vigorously because genomic DNA breaks easily. For better precipitation results, freeze the sample at −20 °C for at least 1 h, but overnight is recommended.
15. After discarding the supernatant, centrifuge at full speed for an additional 1 min to remove residual solvent and allow the DNA pellet to dry faster.
16. To get a good yield for each of the flanks and marker gene, perform the PCRs in three 50 μL reaction.
17. To maximize the yields of high-quality DNA from the gel extraction process: (i) Run 3–4 identical PCRs and mix the products while loading them by taping multiple wells of the comb together; (ii) pour a thin gel or remove the excess of gel while cutting the band; (iii) minimize exposure to UV light; otherwise, you risk that the DNA exposed to UV for the longest time will be nicked to shreds; (iv) add an extra washing step with the ethanol-containing wash buffer provided in the kit to get rid of chaotropic salt residues on the membrane; and (v) run another centrifuge after the washing steps to ensure all of the ethanol is gone.
18. During the first hour of incubation, gently aspire the suspension up and down, 4–6 times, using a 1 mL pipette to break the hyphal tissue and release protoplasts.
19. In addition to the transformation reaction, include both positive and negative controls in the experimental design. To run the negative control, mix 100 μL of STC buffer with 100 μL of protoplast and incubate the mixture for 50 min on ice. The positive control helps confirm the quality of the protoplasts. To run a positive control, plate 500 μL to 1 mL of the transformation reaction on SMM (not supplemented with selection marker). The fungus should massively grow on the positive control plate. The negative control helps confirm that any positive experimental data is arising from the experimental sample and not from random event.
20. To validate that the absence of amplification band is due to deletion of the target sequence in the template DNA and not to other reasons such as DNA quality or concentration, conduct a housekeeping gene amplification (e.g., β-actin, β-tubulin, or others).

21. It is recommended to design and perform two southern blotting to validate the integration events on the upstream and downstream side. For that, use the 5′- and 3′- flank purified DNA fragments that served for the construction of the gene deletion cassette for preparing probes for southern blot.
22. The following day, disassemble the Southern blot assembly. Do not incubate it longer as this may result in failure of the blot.
23. Before assembling the Southern blot setup, cut the Hybond membrane and Whatman paper (to place on top of the gel) the same size as the gel. Before overnight incubation, cover the Pyrex glass with saran wrap to minimize evaporation of the transfer buffer.
24. Follow general precautions for working with radioactive material. Use a Geiger counter radiation detector to monitor for contamination. It is recommended to use the P32 within 2 weeks from receiving it for better efficiency.
25. Deepot cells should be used for softwoods, hardwoods, or stalk vegetables, to allow water drainage from the bottom. Pots should be used for stem vegetables at their late development stage so water can be retained on the tray. Cell trays are best to be used with stem vegetables at their early development stage so water can be retained on the tray.
26. It is recommended to conduct a preliminary study to determine how much macronutrients and micronutrients to add without affecting the experimental design.
27. To determine the inoculum size (number of spores) needed to induce disease in plants, conduct a preliminary assay using different spore concentrations.
28. Adjust the volume of water and ethyl acetate for large-scale extraction of secondary metabolites. Every 20 mL of agar medium is ground in 5 mL of ddH_2O and extracted using 5 mL of ethyl acetate.

Acknowledgements

The Secure Ecosystem Engineering and Design (SEED) (https://seed-sfa.ornl.gov/) project is funded by the Genomic Science Program of the US Department of Energy, Office of Science, Office of Biological and Environmental Research (BER), as part of the Secure Biosystems Design Science Focus Area (SFA). All figures were created with http://www.biorender.com/.

References

1. Hyde KD, Xu J, Rapior S et al (2019) The amazing potential of fungi: 50 ways we can exploit fungi industrially. Fungal Divers 97:1–136
2. Calvo AM, Wilson RA, Bok JW, Keller NP (2002) Relationship between secondary metabolism and fungal development. Microbiol Mol Biol Rev 66:447–459
3. Drott MT, Debenport T, Higgins SA et al (2019) Fitness cost of aflatoxin production in *Aspergillus flavus* when competing with soil microbes could maintain balancing selection. MBio 10:e02782–e02718
4. Dufour N, Rao RP (2011) Secondary metabolites and other small molecules as intercellular pathogenic signals. FEMS Microbiol Lett 314:10–17
5. Netzker T, Fischer J, Weber J et al (2015) Microbial communication leading to the activation of silent fungal secondary metabolite gene clusters. Front Microbiol 6:1–13
6. Kai K (2019) Bioorganic chemistry of signaling molecules in microbial communication. J Pestic Sci 44:200–207
7. Scharf DH, Heinekamp T, Brakhage AA (2014) Human and plant fungal pathogens: the role of secondary metabolites. PLoS Pathog 10:10–12
8. Pusztahelyi T, Holb IJ, Pócsi I (2015) Secondary metabolites in fungus-plant interactions. Front Plant Sci 6:1–23
9. Keller NP (2019) Fungal secondary metabolism: regulation, function and drug discovery. Nat Rev Microbiol 17:167–180
10. Gacek A, Strauss J (2012) The chromatin code of fungal secondary metabolite gene clusters. Appl Microbiol Biotechnol 95:1389–1404
11. Keller NP, Turner G, Bennett JW (2005) Fungal secondary metabolism – from biochemistry to genomics. Nat Rev Microbiol 3:937–947
12. Chavali AK, Rhee SY (2018) Bioinformatics tools for the identification of gene clusters that biosynthesize specialized metabolites. Brief Bioinform 19:1022–1034
13. Brakhage AA, Schroeckh V (2011) Fungal secondary metabolites – strategies to activate silent gene clusters. Fungal Genet Biol 48:15–22
14. Möbius N, Hertweck C (2009) Fungal phytotoxins as mediators of virulence. Curr Opin Plant Biol 12:390–398
15. Marasas WFO, Riley RT, Hendricks KA et al (2004) Fumonisins disrupt sphingolipid metabolism, folate transport, and neural tube development in embryo culture and in vivo: a potential risk factor for human neural tube defects among populations consuming fumonisin-contaminated maize. J Nutr 134:711–716
16. Abbas HK, Tanaka T, Duke SO, Porter JK, Wray EM, Hodges L, Sessions AE, Wang E, Merrill AH Jr, Riley RT (1994) Fumonisin- and AAL-toxin-induced disruption of sphingolipid metabolism with accumulation of free sphingoid bases. Plant Physiol 106:1085–1093
17. Meena M, Samal S (2019) *Alternaria* host-specific (HSTs) toxins: an overview of chemical characterization, target sites, regulation and their toxic effects. Toxicol Rep 6:745–758
18. Turgeon BG, Baker SE (2007) Genetic and genomic dissection of the *Cochliobolus heterostrophus* Tox1 locus controlling biosynthesis of the polyketide virulence factor T-toxin. Adv Genet 57:219–261
19. Luo Z, Ren H, Mousa JJ et al (2017) The PacC transcription factor regulates secondary metabolite production and stress response, but has only minor effects on virulence in the insect pathogenic fungus *Beauveria bassiana*. Environ Microbiol 19:788–802
20. Barda O, Maor U, Sadhasivam S et al (2020) The pH-responsive transcription factor PacC governs pathogenicity and ochratoxin A biosynthesis in *Aspergillus carbonarius*. Front Microbiol 11:1–11
21. Tannous J, Kumar D, Sela N et al (2018) Fungal attack and host defence pathways unveiled in near-avirulent interactions of *Penicillium expansum creA* mutants on apples. Mol Plant Pathol 19:2635–2650
22. Fasoyin OE, Wang B, Qiu M et al (2018) Carbon catabolite repression gene *creA* regulates morphology, aflatoxin biosynthesis and virulence in *Aspergillus flavus*. Fungal Genet Biol 115:41–51
23. Giese H, Sondergaard TE, Sørensen JL (2013) The AreA transcription factor in *Fusarium graminearum* regulates the use of some nonpreferred nitrogen sources and secondary metabolite production. Fungal Biol 117:814–821
24. Pfannmüller A, Leufken J, Studt L et al (2017) Comparative transcriptome and proteome analysis reveals a global impact of the nitrogen regulators AreA and AreB on secondary metabolism in *Fusarium fujikuroi*. PLoS One 12:1–27

25. Bok JW, Keller NP (2004) LaeA, a regulator of secondary metabolism in *Aspergillus spp.* Eukaryot Cell 3:527–535
26. Lind AL, Smith TD, Saterlee T et al (2016) Regulation of secondary metabolism by the velvet complex is temperature-responsive in *Aspergillus*. G3-Genes Genom Genet 6:4023–4033
27. Agrios G (2004) Plant pathology – 5th edition. Academic Press
28. Kukurba KR, Montgomery SB (2015) RNA sequencing and analysis. Cold Spring Harb Protoc 2015:951–969
29. Corchete LA, Rojas EA, Alonso-López D et al (2020) Systematic comparison and assessment of RNA-seq procedures for gene expression quantitative analysis. Sci Rep 10:1–15
30. Li WV, Li JJ (2018) Modeling and analysis of RNA-seq data: a review from a statistical perspective. Quant Biol 6:195–209
31. Blin K, Shaw S, Steinke K et al (2019) anti-SMASH 5.0: updates to the secondary metabolite genome mining pipeline. Nucleic Acids Res 47:W81–W87
32. Jung H, Ventura T, Sook Chung JS et al (2020) Twelve quick steps for genome assembly and annotation in the classroom. PLoS Comput Biol 16:1–25
33. Greco C, Pfannenstiel BT, Liu JC, Keller NP (2019) Depsipeptide aspergillicins revealed by chromatin reader protein deletion. ACS Chem Biol 14:1121–1128
34. Tannous J, Snini SP, El Khoury R et al (2017) Patulin transformation products and last intermediates in its biosynthetic pathway, E- and Z-ascladiol, are not toxic to human cells. Arch Toxicol 91:2455–2467

Chapter 16

Eliciting Targeted Mutations in *Medicago sativa* Using CRISPR/Cas9-Mediated Genome Editing: A Potential Tool for the Improvement of Disease Resistance

Udaya Subedi, Kimberley Burton Hughes, Guanqun Chen, Abdelali Hannoufa, and Stacy D. Singer

Abstract

CRISPR/Cas9 (clustered regularly interspaced short palindromic repeats/CRISPR-associated protein 9) has become a breeding tool of choice for eliciting targeted genetic alterations in crop species as a means of improving a wide range of agronomic traits, including disease resistance, in recent years. With the recent development of CRISPR/Cas9 technology in *Medicago sativa* (alfalfa), which is an important perennial forage legume grown worldwide, its use for the enhancement of pathogen resistance is almost certainly on the horizon. In this chapter, we present detailed procedures for the generation of a single nonhomologous end-joining-derived indel at a precise genomic locus of alfalfa via CRISPR/Cas9. This method encompasses crucial steps in this process, including guide RNA design, binary CRISPR vector construction, *Agrobacterium*-mediated transformation of alfalfa explants, and molecular assessments of transformed genotypes for transgene and edit identification.

Key words Alfalfa, CRISPR/Cas, Guide RNAs, Non-homologous end-joining, Targeted mutations

1 Introduction

Alfalfa (*Medicago sativa* L.) is a perennial leguminous forage crop grown worldwide [1] that possesses many positive attributes, including relatively high nutritional value, palatability, and adaptability to various growth conditions [2]. However, alfalfa production can be severely hampered by several diseases caused by bacterial and fungal pathogens affecting the roots or foliage, all of which compromise biomass yield and quality [3–6]. Of these, bacterial wilt (*Clavibacter michiganensis* subsp. *insidiosus*), Verticillium wilt (*Verticillium alfalfae*), *Fusarium* wilt (*Fusarium oxysporum*), and Phytophthora root rot (*Phytophthora medicaginis*), as well as the foliar disease anthracnose (*Colletotrichum trifolii*), tend to be the most problematic [7]. Control measures for many of these diseases

Nora A. Foroud and Jonathan A. D. Neilson (eds.), *Plant-Pathogen Interactions*, Methods in Molecular Biology, vol. 2659, https://doi.org/10.1007/978-1-0716-3159-1_16,

typically consist of either chemical sprays, seed treatments, or soil drenches, which are not ideal from the perspective of human health, cost for the producer, or the environment [5]. Furthermore, they often become only partially effective over time due to the development of resistance by the pathogen [4, 6]. While conventional breeding methods have also been used to successfully develop alfalfa varieties with some level of resistance to certain pathogens [8], further improvements in this area are much needed.

Despite some successes over the years, conventional breeding approaches are complicated by alfalfa's outcrossing and autotetraploid ($2n = 4 \times = 32$) nature [9]. In order to overcome these obstacles, genetic engineering through the overexpression of endogenous or heterologous genes, or the downregulation of target genes, has been the focus of much research for more than two decades. While many positive outcomes have been reported using such methods [10–14], exceedingly stringent regulatory requirements and negative public perception have impeded the commercialization of "genetically modified" ("GM") crops in general.

The use of CRISPR/Cas (clustered regularly interspaced short palindromic repeats/CRISPR-associated protein)-mediated genome editing has also been gaining momentum in recent years as a means of improving the pace and precision of breeding [15–17]. This tool is especially valuable in polyploid species due to its ability to provoke a range of heterozygous, monoalleleic, and biallelic mutations in a single generation [18–20], which is typically not possible with conventional breeding approaches. While various permutations of the CRISPR/Cas platform currently exist [21–25], in its simplest form, the technology depends upon a small customizable guide RNA (gRNA) that directs the Cas nuclease to a selected genomic locus immediately upstream of a protospacer adjacent motif (PAM), where it elicits a double-stranded DNA break (DSB) [26]. Cas9 from *Streptomyces pyogenes* is the most widely used Cas protein for genome editing purposes in plants and requires a PAM of 5′– NGG – 3′, or less frequently 5′–NAG – 3′ [27, 28], with the DSB occurring approximately three nucleotides upstream of the PAM [27]. The DSB that is induced by Cas9 is then repaired by the plant cell's endogenous repair machinery.

The principal route used in the context of genome editing is the error-prone nonhomologous end-joining repair (NHEJ) mechanism, which characteristically leads to the generation of small insertions or deletions (indels) that disrupt the targeted gene. In the context of improving disease resistance, such an approach is only in its infancy as of yet and has typically focused on the knockout of genes with negative regulatory functions in disease response, such as *EUKARYOTIC TRANSLATION INITIATION FACTOR 4E* (*EIF4E*) in Arabidopsis, *OsSWEET11* in rice, and *MITOGEN-ACTIVATED PROTEIN KINASE 3* (*SlMAPK3*) in tomato [29–

31]. While this strategy has shown promise in many crop species to date [15, 16, 32–34], it has yet to be successfully implemented in alfalfa. However, CRISPR/Cas9 has been shown to be an effective breeding tool in alfalfa recently, yielding plants with altered phenotypes in the first generation [19, 20]. In addition, although regulatory policies surrounding the use of CRISPR/Cas technology for crop improvement have yet to be finalized in many countries, with the exception of the European Union and New Zealand, the trend thus far minimally supports the "non-GM" status of plants bearing a simple NHEJ-derived mutation achieved through the use of a site-specific nuclease system such as CRISPR/Cas [26]. This stems mainly from the fact that the resulting mutations in such cases are indistinguishable from those occurring spontaneously or from conventional breeding techniques [35]. As such, there is little doubt that CRISPR/Cas will provide a novel breeding tool with the potential to enable further improvements in various traits, including pathogen resistance, in alfalfa.

In this chapter, we describe a basic method for achieving a single NHEJ-based indel at a precise genomic locus using CRISPR/Cas9 in alfalfa, a method that could potentially be used to target any gene of choice, including suitable genes for eliciting disease resistance in this species. Briefly, our method involves the design of a binary CRISPR vector based on the pKSE401 system [36] and the *Agrobacterium*-mediated delivery of the construct into an amenable alfalfa genotype, as well as molecular assessments of transformed genotypes via PCR and detection of edits using T7 endonuclease1 (T7E1) assays, droplet digital PCR (ddPCR), and Sanger sequencing (Fig. 1).

2 Materials

All solutions should be prepared using UltraPure water and can be stored at room temperature unless otherwise specified. In addition to the specific materials listed below, this protocol also requires standard molecular biology/greenhouse supplies and consumables, including 1.5 mL microcentrifuge tubes, Falcon tubes of various sizes, PCR tubes, agarose gel systems and associated supplies, thermocycler, Bunsen burner or equivalent, centrifuges, heating block and/or heated water bath, shaking incubator, petri plates, Luria-Bertani (LB) medium, agar, sterilization equipment/supplies, parafilm, 70% ethanol, pipets and an assortment of tips, approximately 10 cm square pots, flat trays, appropriate potting mix, and clear domes for covering the pots.

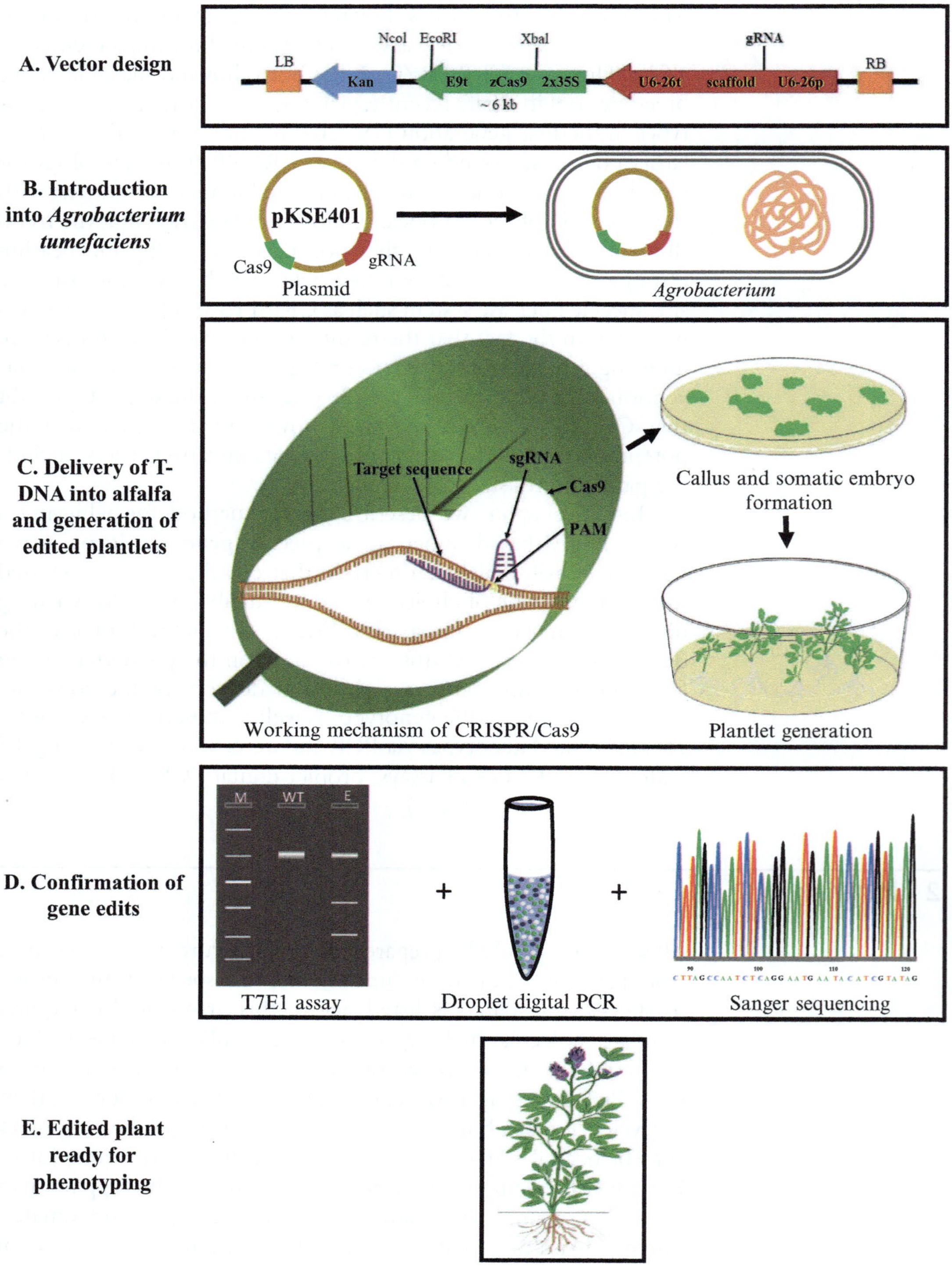

Fig. 1 Key steps in the generation of CRISPR/Cas9-mediated NHEJ-based edits in alfalfa. (**a**) Vector design. LB, left border; Kan, kanamycin resistance cassette; E9t, pea RuBisCO small subunit E9 terminator; zCas9, *Zea mays* codon-optimized Cas9; 2x35S, 2× CaMV35S promoter; U6–26t, Arabidopsis U6–26 gene terminator; U6–26p, Arabidopsis U6–26 gene promoter; *RB* right border. (**b**) Cloning of binary editing vector and introduction into *Agrobacterium tumefaciens*. (**c**) Delivery of T-DNA into the host system and generation of transformed plantlets. (**d**) Confirmation of gene edits. Genomic DNA (isolated from genotypes confirmed to be

2.1 Construct Assembly

1. User-defined gRNA oligonucleotides: Dilute oligonucleotides to 100 μM with sterile, nuclease-free water and store at −20 °C.
2. Vector pKSE401 (*see* **Note 1**). Store at −20 °C.
3. BsaI-HF restriction enzyme. Store at −20 °C (*see* **Note 2**).
4. Gel DNA extraction kit of choice.
5. T4 DNA ligase with associated buffer. Store at −20 °C.
6. Chemically competent *Escherichia coli*. Store at −80 °C.
7. 50 mg/mL kanamycin stock: Dissolve 500 mg kanamycin in 10 mL water. Filter sterilize and store in 1.5 mL aliquots at −20 °C.
8. LB-Kan solid medium: Mix 1 L LB broth (using pre-mixed powder or separate components), and then add 15 g agar. Autoclave, cool to approximately 50 °C, then add 1 mL 50 mg/mL kanamycin (final concentration of 50 mg/L), and pour into petri plates. Once solidified, store at 4 °C.
9. Primers for colony PCR: U6–26p-F (5′ – TGT CCC AGG ATT AGA ATG ATT AGG C – 3′) and U6–26t-R (5′ – CCC CAG AAA TTG AAC GCC GAA GAA C – 3′) [36].
10. DNA polymerase master mix of choice.
11. LB-Kan liquid medium: Mix 1 L LB broth (using pre-mixed powder or separate components), autoclave, and then add 1 mL 50 mg/mL kanamycin (final concentration of 50 mg/L) once cool. Store at 4 °C.
12. Plasmid extraction mini prep kit of choice.

2.2 *Agrobacterium*-Mediated Transformation of Alfalfa Leaf and Petiole Explants

1. Electrocompetent *Agrobacterium tumefaciens* LBA4404 cells (*see* **Note 3**). Store in 50 μL aliquots at −80 °C.
2. Electroporator.
3. 0.1 cm electroporation cuvettes.
4. 50 mg/mL rifampicin stock: Dissolve 250 mg in 5 mL DMSO. Filter sterilize and store in 0.5 mL aliquots at −20 °C (*see* **Note 4**).
5. 50 mg/mL kanamycin stock.
6. LB-Rif/Kan solid medium: Prepare as for LB-Kan solid medium. Add 1 mL/L of 50 mg/mL rifampicin (final concentration of 50 mg/L) before pouring into the plates.

Fig. 1 (continued) transgenic via PCR) is subjected to a T7 endonuclease 1 (T7E1) assay and droplet digital PCR to confirm mutation events at the targeted site. Sanger sequencing is then used to determine the precise nature of the edit in edited genotypes and to detect any off-target mutations at the highest likelihood sites. *M* gel marker, *WT* wild type, *E* genome edited line. (**e**) Edited genotype that is ready for phenotyping

7. LB-Rif/Kan liquid medium: Prepare as for LB-Kan liquid medium. Add 1 mL/L of 50 mg/mL rifampicin (final concentration of 50 mg/L).
8. 60% glycerol: Mix 60 mL glycerol with 40 mL water. Autoclave.
9. 100 mM acetosyringone stock. Order from supplier, aliquot into 1.5 mL tubes, and store at −20 °C.
10. LB-acetosyringone/Rif/Kan liquid medium: Prepare as for LB-Rif/Kan liquid medium. Add 200 μL/L of 100 mM acetosyringone (final concentration of 20 μM) immediately prior to using the media.
11. 10 mg/mL pyridoxine HCl: Dissolve 100 mg pyridoxine HCl in 10 mL water, filter sterilize, and store at 4 °C.
12. 10 mg/mL thiamine HCl: Dissolve 100 mg thiamine HCl in 10 mL water, filter sterilize, and store at 4 °C.
13. 10 mg/mL glycine: Dissolve 100 mg glycine in 10 mL water, filter sterilize, and store at 4 °C.
14. 25 mg/mL thioproline: Dissolve 1 g thioproline in 40 mL water, filter sterilize, and store at 4 °C.
15. 300 mg/mL Timentin stock: Dissolve 3 g Timentin in 10 mL of water, filter sterilize, aliquot into 1.5 mL tubes, and store at −20 °C.
16. 10× Schenk and Hildebrandt (SH)-modified vitamins with myo-inositol: Add 50 mg nicotinic acid, 0.5 mL of 10 mg/mL pyridoxine HCl, 50 mg thiamine HCl, and 2 g myo-inositol in a beaker, and make up to 1 L with water. Filter sterilize and store at 4 °C.
17. 10× Blade's stock with myo-inositol: Add 350 mg of $MgSO_4.7H_20$, 44 mg of $MnSO_4.H_2O$, 3.47 g of $Ca(NO_3)_2.4H_20$, 10 g of NH_4NO_3, 10 g of KNO_3, 3 g of KH_2PO_4, 650 mg of KCl, 16 mg of H_3BO_3, 15 mg of $ZnSO_4.7H_2O$, 8 mg of KI, 360 mg of Fe(III)EDTA, 5 mg of nicotinic acid, 0.1 mL of 10 mg/mL pyridoxine HCl, 100 μL of 10 mg/mL thiamine HCl, 2 mL of 10 mg/mL glycine, and 1 g of myo-inositol in a beaker, and make up to 1 L with water. Filter sterilize and store at 4 °C.
18. 10× Murashige and Skoog (MS)-modified vitamins: Add 5 mg nicotinic acid, 0.5 mL of 10 mg/mL pyridoxine HCl, 1 mL of 10 mg/ml thiamine HCl, and 2 g myo-inositol in a beaker, and make up to 1 L with water. Filter sterilize and store at 4 °C.
19. Basal SH2K medium: Add 3.2 g of Schenk and Hildebrandt salts, 100 mL of 10× SH -modified vitamins, 4.35 g of K_2SO_4, 288 mg of proline, and 30 g of sucrose in a beaker, and then add 900 mL of water. Adjust pH to 5.8 and then make up the

solution to 1 L. Autoclave, cool to approximately 50 °C, and add 40 μL of 10 mg/mL kinetin and 400 μL of 10 mg/mL 2,4-D prior to plating.

20. 100 × 25 mm deep petri plates.
21. 120 × 80 mm tissue culture vessels with lids (*see* **Note 5**).
22. Preculture (PC) liquid medium: 1 L basal SH2K medium with 2.12 mL 25 mg/mL thioproline (final concentration of 53 mg/L).
23. PC solid medium: 1 L PC liquid medium. Add 8 g agar prior to autoclaving.
24. Co-cultivation (CO) liquid medium: 1 L PC liquid medium containing 200 μL/L of 100 mM acetosyringone stock solution (final concentration of 20 μM).
25. CO solid medium: 1 L CO liquid medium. Add 8 g agar prior to autoclaving.
26. Callus induction (CI) medium: 1 L PC liquid medium. Add 8 g agar prior to autoclaving and 1 mL 300 mg/mL Timentin after autoclaving (final concentration of 300 mg/L).
27. CI selection medium-1: 1 L CI medium containing 1 mL 50 mg/mL kanamycin (added after autoclaving; final concentration of 50 mg/L).
28. CI selection medium-2: 1 L CI medium containing 1.5 mL 50 mg/mL kanamycin (added after autoclaving; final concentration of 75 mg/L).
29. Embryo induction (EI) medium: 100 mL 10× Blade's stock with myo-inositol, 2 g yeast extract, and 30 g sucrose; make up to 1 L with water, pH to 5.8 with HCl. Add 8 g agar prior to autoclaving. Add 1 mL 300 mg/mL Timentin (final concentration of 300 mg/L) and 1.5 mL 50 mg/mL kanamycin (final concentration of 75 mg/L) after autoclaving once media has cooled to approximately 50 °C.
30. Embryo germination (EG) medium: 2.165 g MS basal salts (no vitamins), 100 ml 10× MS-modified vitamins, 100 μL 10 mg/L glycine, and 30 g sucrose; make up to 1 L with purified water, pH 5.8, with KOH. Add 8 g agar prior to autoclaving. Add 1 mL 300 mg/mL Timentin (final concentration of 300 mg/L) and 1.5 mL 50 mg/mL kanamycin (final concentration of 75 mg/L) after autoclaving once media has cooled to approximately 50 °C. Pour into 100 × 25 mm petri plates.
31. Plant development (PD) medium: 4.33 g MS basal salts (no vitamins), 100 mL 10× MS-modified vitamins, 100 μL 10 mg/L glycine, and 30 g sucrose; make up to 1 L with purified water, pH to 5.8, with KOH. Add 8 g agar prior to autoclaving. Add

1 mL 300 mg/mL Timentin (final concentration of 300 mg/L) and 1.5 mL 50 mg/mL kanamycin (final concentration of 75 mg/L) after autoclaving once media has cooled to approximately 50 °C. Pour into containers.

32. Spectrophotometer.
33. Alfalfa genotype amenable to regeneration in tissue culture (e.g., N.4.4.2 [38] grown in greenhouse or growth cabinet).
34. Laminar flow hood.
35. 70% ethanol in a 50 mL tube.
36. Leaf sterilization solution: 20% bleach with one drop of liquid dish soap in 50 mL tube.
37. Sterile water in 50 mL tubes (three tubes per transformation).
38. Forceps.
39. Sterile shark skin filter paper: Place two sterile sheets into a petri plate (upside down) and moisten with sterile water (*see* **Note 6**). You will require several plates per transformation.
40. Razor blades and holders.
41. Micropore tape.
42. Tissue culture incubator/cabinet.
43. Sterile Whatman filter paper: Place two sterile sheets into a petri plate (upside down). You will require several plates per transformation.

2.3 PCR Validation of Transgenic Genotypes

1. Kit or method for extracting genomic DNA.
2. DNA polymerase master mix of choice.
3. Primers for assessing the presence of the Cas9 cassette: We use Cas9F1 (5′ – GCT GGA GGA GTC ATT CCT CG – 3′) and Cas9R1 (5′ – CTG AGA ATA TCA GAC AGG AGG – 3′).

2.4 Identification of Edited Genotypes

2.4.1 T7E1 Assays

1. T7 endonuclease I (T7E1; 10,000 U/mL) and associated buffer (e.g., New England Biolabs).
2. DNA polymerase master mix of your choice.
3. User-defined primers: Dilute primers to 10 μM with sterile, nuclease-free water and store at −20 °C.

2.4.2 Droplet Digital PCR

1. Droplet digital PCR system (*see* **Note 7**), along with associated software and consumables (*see* **Note 8**).
2. ddPCR™ Supermix for Probes no UTP (Bio-Rad).
3. 6-Carboxyfluorescein (FAM)-labelled locked nucleic acid (LNA) probe (NHEJ-sensitive probe): Resuspend probe in TE (pH 8.0) to 10 μM. Prepare small working aliquots to avoid multiple freeze-thaw cycles and store at −20 °C.

4. Hexachlorofluorescein (HEX)-labelled probe (NHEJ-insensitive probe): Resuspend probe in TE (pH 8.0) to 10 μM (*see* **Note 9**). Prepare small working aliquots to avoid multiple freeze-thaw cycles and store at −20 °C.
5. User-defined primers that generate an amplicon spanning both probe annealing sites: Dilute primers to 10 μM with sterile, nuclease-free water and store at −20 °C.
6. Synthetic DNA corresponding to ddPCR amplicon with single base pair deletion or insertion to assess the sensitivity of the NHEJ-sensitive probe (*see* **Note 10**).
7. TaqMan probe-based qPCR master mix of choice for testing of primers/probes.

2.4.3 Sanger Sequencing

1. High-fidelity DNA polymerase master mix.
2. User-defined primers for the amplification of regions encompassing the target site, as well as primer pairs for the amplification of a set of possible off-target sites.
3. Gel extraction kit.
4. TA cloning kit, chemically competent *E. coli* appropriate for blue-white screening and media components for blue-white screening.
5. 10 mM dATP.
6. Taq DNA polymerase, with associated 10× PCR buffer and 50 mM magnesium chloride.
7. LB liquid medium with appropriate antibiotic for TA cloning vector.
8. Plasmid extraction mini prep kit of choice.
9. Appropriate restriction enzymes for confirming the presence of the PCR insert.
10. Appropriate primer for sequencing insert.

3 Methods

3.1 Selection of Target Sequence and Guide RNA Design

1. Identify target sequence(s) using Blast searches against the *M. sativa* (cultivated alfalfa at the diploid level) genome sequence (available at www.alfalfatoolbox.org) and/or the Alfalfa Gene Index and Expression Atlas database (available within LegumeIP V3; http://plantgrn.noble.org/LegumeIP/gdp/12/gene). Alternatively, the genome sequence of tetraploid alfalfa can be downloaded from https://figshare.com/articles/dataset/genome_fasta_sequence_and_annotation_files/12327602 [20] and searched using your preferred sequence analysis software (*see* **Note 11**).

2. Select a 20-nt region immediately upstream of a 5′ – NGG – 3′ PAM sequence within an exon for your gRNA (*see* **Notes 12** and **13**).
3. The specificity and potential off-target effects of each gRNA can be assessed using freely available software such as Cas-OFFinder [39].
4. Once your gRNA target sequences have been selected, corresponding sense and antisense oligonucleotides must be generated. In both instances, adapter sequences must be added to the 5′ ends to facilitate cloning into the BsaI site of linearized pKSE401. For the sense oligonucleotide, the sequence 5′ – ATTG – 3′ must be added to the 5′ terminus, while for the antisense oligonucleotide, 5′ – AAAC – 3′ must be added to the 5′ terminus [36] (Fig. 2).

3.2 Construct Assembly

Plant binary constructs are generated as described in [36] with minor modifications. All steps are carried out on ice unless otherwise specified.

1. Digest 3 μg pKSE401 with 1 μL (20 U/μL) BsaI in a volume of 30 μL. Incubate the mixture at 37 °C for 3 h.
2. Add loading dye and run the entire digest on a 1% agarose gel (*see* **Note 14**) alongside an appropriate DNA ladder.
3. Cut out the large band (~15.2 kb) using a sterile razor blade and purify using the gel purification kit of your choice. Elute the final product in 30 μL of provided buffer.
4. Mix 3 μL of the corresponding forward and reverse 100 μM oligonucleotides, respectively, and incubate at 65 °C for 5 min. Allow to cool slowly at room temperature for approximately 5 min to allow for annealing into double-stranded molecules.
5. To insert the annealed gRNAs into the pKSE401 vector, combine 2 μL purified BsaI-digested pKSE401 (large fragment), 2 μL annealed oligonucleotide mixture, 2 μL T4 DNA ligase buffer, and 1 μL T4 DNA ligase in a final volume of 20 μL. Incubate the mixture at room temperature for 1 h.
6. Thaw competent *E. coli* cells on ice and transform with ligation mixture as recommended for the particular bacterial strain used. Plate onto LB-Kan solid medium and incubate overnight at 37 °C.
7. The following day, select five separate colonies and carry out colony PCR according to the DNA polymerase master mix manufacturer's recommendations with primers U6-26p-F and U6-26t-R (*see* **Notes 15** and **16**).

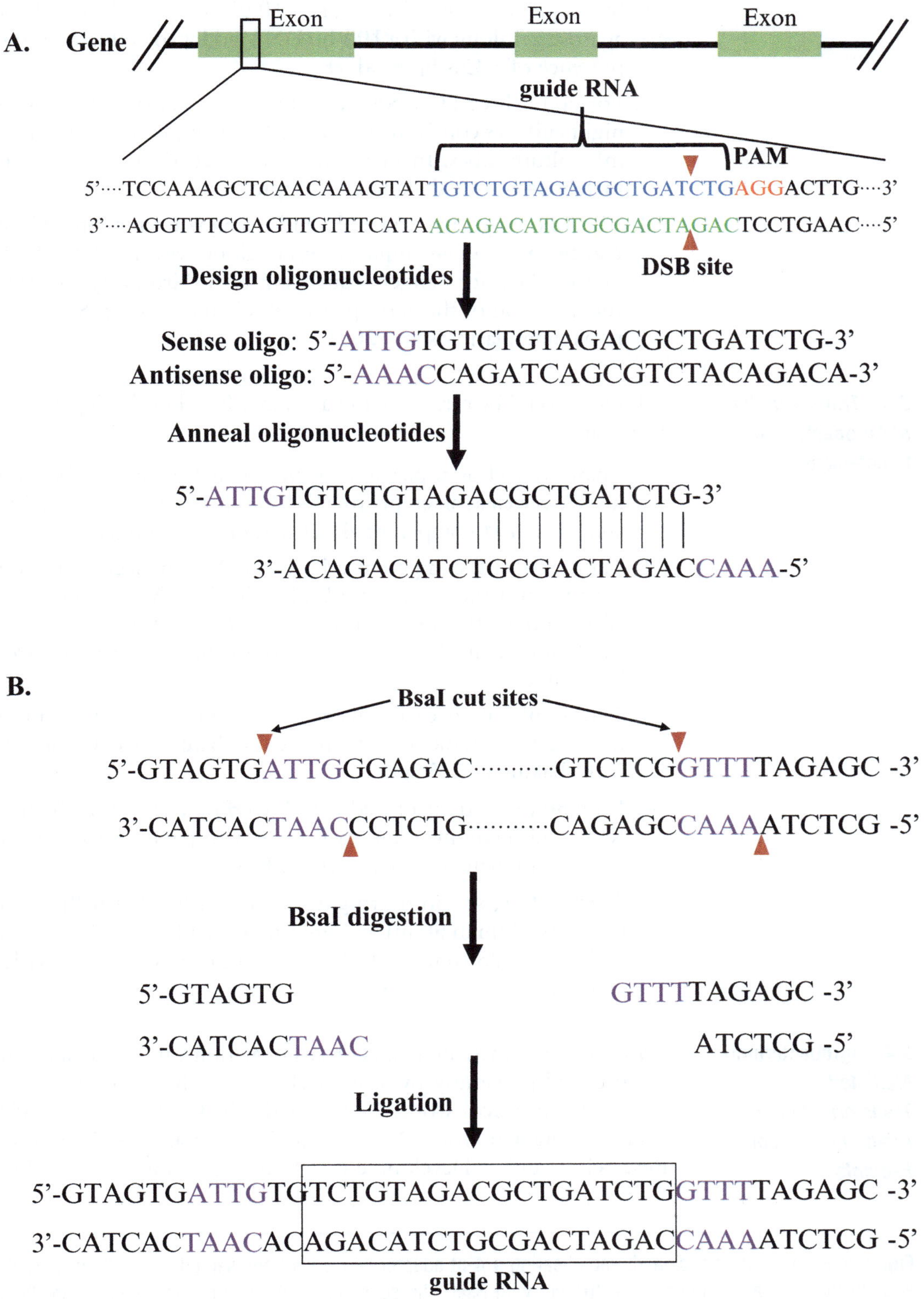

Fig. 2 Example of the introduction of a user-defined gRNA into the pKSE401 vector. (**a**) A 20-nt sequence immediately upstream of a 5′-NGG-3′ PAM and within an exon near the 5′ terminus of the target gene is selected for gRNA design. Sense and antisense oligonucleotides corresponding to this 20-nt sequence are

8. Subsequent to amplification, run 10 μL of the PCR on a 1.5% agarose gel along with a 100-bp DNA ladder to evaluate for the presence of a 423-bp product.
9. For up to three PCR-positive colonies per vector, set up overnight cultures consisting of 3 mL LB-Kan liquid medium in 15 mL culture tubes. Incubate the cultures at 37 °C for approximately 24 h with shaking at 250 rpm.
10. The following day, extract plasmids using your mini prep kit of choice and send an aliquot of each, along with the U6-26p-F primer [36], for Sanger sequencing by a third party to verify the inclusion of the appropriate gRNA in each case. Store the remaining plasmids at −20 °C.

3.3 Transformation of *Agrobacterium tumefaciens*

All steps should be carried out in a laminar flow hood using aseptic technique.

1. Thaw your chosen plasmids (experimental plasmid(s) bearing gRNA as well as an empty pKSE401 control bearing no gRNA) and LBA4404 competent *Agrobacterium* cells on ice.
2. Transfer 40 μL competent cells into a 0.1 cm electroporation cuvette, and then add 0.5 μL plasmid directly into the cells. Electroporate the mixture using an electroporator according to the manufacturer's instructions. Immediately place cuvettes back on ice.
3. Add 1 ml LB to each cuvette, and then transfer the entire mixture to a sterile 1.5 mL tube. Incubate at 28 °C for 3 h with shaking at 225 rpm.
4. Plate 50 μL of transformed *Agrobacterium* cells onto LB-Rif/Kan solid medium. Wrap each plate with parafilm to prevent drying and incubate for 2 days at 28 °C.
5. After 2 days, set up overnight cultures in 5 mL LB-Rif/Kan liquid medium in 50 mL culture tubes. Incubate at 28 °C with shaking at 225 rpm for 24 h. Store cultures as glycerol stocks (30% glycerol) at −80 °C.

3.4 *Agrobacterium*-Mediated Transformation of Alfalfa Leaf/Petiole Explants

Agrobacterium-mediated alfalfa transformation can be carried out as described previously by Tian et al. [31] with minor modifications. This protocol works well with the N.4.4.2 genotype [38] (can be obtained from Dr. Abdelali Hannoufa, AAFC London Research and Development Centre); however, other

Fig. 2 (continued) then designed, each bearing a 4-nt adapter sequence (represented in purple) at their 5′ ends to facilitate annealing to pKSE401 cut with BsaI. The sense and antisense oligonucleotides are then annealed to form a double-stranded molecule. *DSB* double-stranded break, *PAM* protospacer adjacent motif. (**b**) The pKSE401 vector bears two BsaI sites, between which the double-stranded gRNA is ligated to yield the final binary vector

transformation protocols can also be used with other genotypes if desired. All steps should be carried out in a laminar flow hood using aseptic technique. The tissue culture cabinet was maintained at 26 ° C and 75 μmol m^{-2} s^{-1} light intensity throughout.

1. On day 1, set up 5 mL *Agrobacterium* starter cultures from glycerol stocks in LB-Rif/Kan liquid medium in a sterile 50 mL tube. Incubate for approximately 24 h at 28 °C with shaking at 225 rpm.
2. Harvest approximately 20 healthy trifoliate leaves (including petioles) from the N.4.4.2 alfalfa genotype and place immediately in a 50 mL tube containing water. This step can be carried out in the greenhouse.
3. Sterilize trifoliate leaves by first placing approximately 7–10 leaves in 70% ethanol for 10 s, then transferring to leaf sterilization solution for 5 min, and finally rinsing three times in sterile water (in separate 50 mL tubes). Leave the leaves in the final tube of water, and repeat until all tissue has been sterilized (*see* **Note 17**).
4. Transfer approximately 4–5 leaves (using sterile forceps) from water onto sterile, moistened shark skin filter paper. Use a sterile razor (change regularly to ensure it remains sharp) and cut leaflets into quarters (or sixths if leaves are large enough). Petioles can be cut into segments 5–10 mm in length (*see* **Note 18**). Place explants flat on PC plates. Repeat until all tissue has been cut and plated (*see* **Note 19**).
5. Wrap plates twice with micropore tape and place in tissue culture cabinet for 2 days. Cover the plates with a single layer of paper towel to prevent exposure to direct light.
6. On day 2, set up 25 mL cultures in 125 mL flasks with LB-acetosyringone/Rif/Kan liquid medium and 250 μL starter culture. Incubate at 28 °C overnight with shaking at 225 rpm.
7. On day 3, measure the OD600 of your culture and dilute to an absorbance of 1.0. Transfer the culture into 2× 50 mL sterile tubes and centrifuge at 3500 rpm for 30 min at 15 °C. Pour off the medium and resuspend the bacterial pellets in 25 mL CO liquid medium.
8. Divide each *Agrobacterium* suspension into two petri plates. Transfer alfalfa explants from PC medium to the bacterial suspensions and allow to soak for 10 min. Ensure the explants are covered in suspension.
9. Following incubation, transfer the explants onto dry sterile filter paper in a petri plate and blot to remove excess liquid. Immediately transfer to CO plates (*see* **Note 20**). Wrap plates with micropore tape, wrap stacks of plates in tin foil, and incubate for 4–5 days (*see* **Note 21**).

10. After 4–5 days, wash explant twice in 25 mL sterile water in 50 mL tubes (two separate tubes) and then once in 25 mL PC liquid medium in a 50 mL tube. Approximately half of the explants from each transformation can be washed at once. Use fresh water and PC medium for each batch.
11. Following washing, blot tissue on sterile filter paper in a petri plate and immediately transfer explants to CI plates (*see* **Note 22**). Wrap twice with micropore tape and incubate in a tissue culture cabinet for 2 weeks. Cover the plates with a single layer of paper towel.
12. After 2 weeks, transfer calli to CI selection medium 1 using sterile forceps and incubate as above for 10 days. Plates do not require covering from this step onward.
13. After 10 days, transfer calli to CI selection medium 2 and incubate as above for a further 1 week.
14. After 1 week, transfer calli to EI medium and incubate for 4–8 weeks. Transfer to fresh EI medium every 2 weeks.
15. Once dark green elongated embryos are evident, transfer individual embryos to EG medium and incubate as above for 2–3 weeks.
16. Once embryos have developed small shoots, transfer the germinated embryos to PD medium and incubate as above for 3–4 weeks.
17. Once roots have formed, transfer plantlets to 15 mL tubes bearing approximately 7 mL of tap water (roots should be in the water and shoots exposed to air within the tube). This step does not require sterile conditions. Incubate on a bench top at room temperature with the lids off for 3 days.
18. After 3 days, transfer each plantlet to moist soil-less medium in approximately 10 cm square pots in the greenhouse or growth cabinet (*see* **Note 23**). Cover the pots with a clear plastic dome for 3–4 days, and then open vents for 3–4 days. At this point, you can partially remove the domes for 2 days, and then remove completely.

3.5 PCR Validation of Transgenic Genotypes

All steps should be carried out on ice unless otherwise specified.

1. Harvest a single leaflet from each transformed genotype, as well as an untransformed plant, and extract genomic DNA using your kit or method of choice.
2. Using the extracted genomic DNA as template, set up PCR assays using your DNA polymerase of choice according to the manufacturer's recommendations, along with primers Cas9F1 and Cas9R1 (*see* **Notes 24** and **25**).

3. Run 10 μL of each product on a 1% agarose gel alongside an appropriately sized DNA ladder to confirm the presence of a 621-bp amplicon. Plants producing the amplicon are transgenic.

3.6 Identification of Edited Genotypes

All steps should be carried out on ice unless otherwise specified.

3.6.1 T7E1 Assays

This assay is based on the ability of T7E1 to specifically cleave heteroduplex DNA at the site of a mismatch (in this case, the edit).

1. Design primers to encompass the predicted site of editing (*see* **Note 26**). Ensure that the edit site is off-center in order to guarantee that if a mismatch is detected (and thus cut) by T7E1, the two resulting DNA fragments can be differentiated on an agarose gel. Also confirm that no other polymorphisms exist among alleles within the amplified region.
2. Use your DNA polymerase master mix of choice to amplify the target region from transgenic DNA samples. Use a no template control, as well as DNA from an untransformed plant and from those bearing empty vector, as negative controls.
3. Confirm the amplification product on an agarose gel and then transfer 10 μL of each PCR sample into two new 0.2 mL PCR tubes.
4. Add 2 μL 10× NEB2 buffer and 7 μL water to each tube.
5. In a thermocycler, heat the tubes to 95 °C for 10 min to denature, then slowly decrease the temperature to 85 °C at a ramp speed of 2 °C/s, and finally decrease the temperature to 25 °C at a ramp rate of 0.3 °C/s.
6. Add 1 μL T7E1 enzyme to one tube of each sample, tap gently to mix, and spin down briefly. Incubate all samples at 37 °C for 15 min.
7. Run 10 μL of each PCR-T7E1 product on a 2% agarose gel (*see* **Note 27**). Samples yielding two bands of the expected size assuming the amplicon is cut at the edit site (often along with a third band derived from uncut amplicon) are considered to be edited, while those with a single band are not.

3.6.2 Droplet Digital PCR

This protocol is derived from methods published previously [40, 41] with modifications and allows the quantification of editing (i.e., the proportion of amplicons that are edited in a given DNA sample).

1. Design a locked nucleic acid (LNA) probe of 10–14 nucleotides (T_m of approximately 66 °C) with a 5' FAM modification and 3' Iowa Black dark quencher (*see* **Notes 28** and **29**). The 5′ region of the probe should anneal to the site where an edit is

expected to occur. This probe should not bind if gene editing has occurred at the probe site and is referred to as the NHEJ-sensitive probe.

2. Design a second probe (less than 30 nt in length) in a conserved region of the gene not expected to be affected by editing (T_m of approximately 68 °C). The probe should possess a 5' HEX modification and 3' Iowa Black dark quencher. Ensure that the two probes do not overlap (*see* **Notes 30**, **31**, and **32**). This probe is designed to bind to all amplicons and is referred to as the NHEJ-insensitive probe.
3. Design PCR primers (18–24 nt in length) encompassing the two probe sites, with a T_m of approximately 62 °C (*see* **Note 33**).
4. Test the efficacy of the primers and probes using TaqMan-based qPCR with genomic DNA from an untransformed plant to ensure that the primers amplify efficiently and that both probes bind unedited DNA from an untransformed plant.
5. Test the sensitivity of the NHEJ-sensitive probe using ddPCR with ddPCR™ Supermix for Probes (no dUTP) in a 96-well plate. Reactions should comprise 10 μL Supermix, 900 nM of each PCR primer, 250 nM of each probe, and synthetic DNA (e.g., gBlocks) consisting of the amplicon bearing a single bp insertion or deletion at the edit site as template.
6. Prepare ddPCR reactions for gene editing frequency (GEF) detection as described in **step 5**, with the exception of template, which will instead comprise approximately 40 ng of genomic DNA from each T7E1-positive genotype, as well as empty vector control genotypes. A no template control is used as a negative control in every assay. Technical replicates are not required.
7. Carry out ddPCR (including the generation of droplets, thermocycling, and droplet analysis) according to the manufacturer's instructions.
8. When droplet reading is complete, analyze using the associated software to identify the ratio of NHEJ-sensitive to NHEJ-insensitive droplets. If no genome editing is detected, the ratio will be 1. The ratio will decrease with the amount of edited copies in the sample. To calculate GEF, the following formula can be used: GEF (%) = (1 – ratio) × 100 [41].

3.6.3 Sanger Sequencing

Sanger sequencing can be carried out with T7E1- and ddPCR-positive plants. This will allow the precise nature of each edit to be determined.

1. Design PCR primers to amplify a region surrounding the cut site from each gRNA. Primers should also be designed to

amplify regions encompassing sequences with the highest potential for off-target mutations with your particular gRNA. Ensure that the resulting amplicons are less than 700 bp to facilitate one-sided sequencing.

2. Amplify the target site from genomic DNA of all T7E1/ddPCR-positive genotypes with a high-fidelity polymerase and purify the PCR amplicon using a gel extraction kit. Also amplify each fragment from an empty vector control genotype as a negative control. Include a no template control for each primer set.
3. Add dATP to the ends of each amplicon using Taq DNA polymerase. Prepare a 30 μL reaction containing 25.4 μL of the amplicon, 1× PCR buffer, 1.5 mM $MgCl_2$, 0.2 μM dATP, and 1 unit Taq DNA polymerase. Incubate at 72 °C for 20 min.
4. Ligate the PCR fragment into a TA vector of choice and transform into chemically competent *E. coli*.
5. For each amplicon, select 24 separate colonies for each edited plant (along with an empty vector negative control plant) using blue/white screening, set up 3 mL overnight cultures in LB liquid medium with the appropriate antibiotic, and incubate overnight at 37 °C with shaking at 250 rpm.
6. Extract plasmid DNA using a miniprep kit of choice and subject the resulting plasmids to test digests using an appropriate restriction enzyme to ensure the presence of the PCR insert.
7. Send the resulting plasmids for Sanger sequencing by a third party with an appropriate primer.
8. Once you have your sequencing results, perform sequence alignments with a reference sequence using a multiple sequence alignment tool of choice to determine the precise edits in each genotype, and confirm the lack of off-target effects at sites with the highest likelihood of occurrence.

4 Notes

1. Can be purchased from Addgene (https://www.addgene.org/; plasmid #62202).
2. We have found that this enzyme should be purchased fresh immediately prior to vector construction.
3. Electrocompetent cells can be produced in house using standard procedures.
4. Protect from light.
5. We use Veritiv 16-oz polypropylene deli container (catalog number 073467) and polypropylene lid (catalog number 075686).

6. We sterilize our filter paper by autoclaving in glass petri dishes wrapped in tin foil.
7. We use the QX200™ AutoDG Droplet Digital™ PCR System, including the AutoDG™ Droplet Generator and QX200™ Droplet Reader, as well as the associated C1000 Touch™ Thermal Cycler with 96-Deep Well Reaction Module and the PX1 PCR Plate Sealer (Bio-Rad).
8. We use the recommended consumables and software for the AutoDG system, including ddPCR™ 96-well plates, DG32™ Automated Droplet Generator Cartridges, Automated Droplet Generation Oil for Probes, Droplet Reader Oil, pipet tips for the AutoDG system, pipet tip waste bin for the AutoDG system, pierceable PCR plate heat seal, and the QuantaSoft™ software (Bio-Rad).
9. HEX and FAM labeling can be used interchangeably on either type of probe, as long as it is noted during analysis.
10. We design a synthetic DNA molecule (e.g., gBlocks, Integrated DNA Technologies) that spans the PCR amplicon, but bears a single nucleotide insertion or deletion at the expected edit site.
11. Ensure that you have identified the positions of exons and introns as you need to ensure your gRNA target sequence falls within an exon in the genomic DNA.
12. Since we have found editing efficiency to be highly dependent on the particular gRNA, we typically design three separate gRNAs for each target (to be used in three separate CRISPR/Cas vectors) as a means of increasing the likelihood of achieving edits.
13. For gRNAs, regions nearer the 5′ terminus of a gene are superior to those near the 3′ terminus as there is a higher likelihood of negatively impacting gene function. Similarly, the effectiveness of gRNAs can be enhanced by selecting regions containing approximately 40–60% GC content [42] and by avoiding TT and GCC motifs within last four bases located proximally to the PAM [43]. A detailed study on several factors to be considered while designing gRNAs for plant genome editing has been reviewed previously [44].
14. Tape two wells together on the gel comb to yield a well large enough to house the entire digest volume.
15. We typically use a reaction mixture containing 12.5 μL Dream Taq Green 2× PCR master mix and 5.5 μL of water, along with 0.5 μM of each primer, with the following thermocycling regime: initial denaturation at 95 °C for 2 min, 35 cycles of 95 °C for 30 s, 55 °C for 30 s, 72 °C for 30 s, and a final extension at 72 °C for 5 min.

16. For colony PCR, use sterile pipette tips to collect cells from each colony and dip into the tube containing the PCR mix.
17. While the tubes of 70% ethanol and leaf sterilization solution can be reused for different batches of leaves, we recommend using fresh tubes of sterile water for each batch.
18. For each transformation, you will need approximately 140 explants.
19. We typically use around 7 PC plates per transformation.
20. We typically use 7–8 CO plates per transformation.
21. Check the CO plates daily for bacterial growth. The explants should not be overgrown with *Agrobacterium* and transferred to callus induction medium typically after 4–5 days.
22. We typically use eight CI plates per transformation.
23. Our plants are grown under greenhouse conditions with a day/night photoperiod of 16 h/8 h and temperature of approximately 22/14 °C.
24. We always carry out PCR assays using DNA from an untransformed plant as a negative control, as well as a no template control. We also carry out a positive control PCR using primers that amplify a similarly sized region of the alfalfa genome to ensure each DNA sample is of sufficient quality for amplification.
25. For the Cas9-specific PCR, we typically use 2× SuperFi PCR mix (Thermo Fisher Scientific) along with a thermal profile of 98 °C for 30 s, 35 cycles of 98 °C for 10 s, 57 °C for 10 s, 72 °C for 30 s, and a final extension of 72 °C for 5 min.
26. We typically use the Integrated DNA Technologies Primer-Quest Tool (https://www.idtdna.com/Primerquest/Home/Index) for primer design.
27. To optimize the separation of small DNA fragments, use fresh electrophoresis buffer, thin gel combs, and a lower voltage.
28. We use the IDT oligoanalyzer with qPCR settings (https://www.idtdna.com/calc/analyzer) for ddPCR probe design.
29. This probe can also comprise standard oligonucleotides; however, we have used LNA probes with success. Locked nucleic acids should be placed on either side of the expected cut site.
30. For the NHEJ-insensitive probe, we typically use a standard oligonucleotide probe; however, an LNA probe with the desired T_m is also acceptable.
31. Ideally, neither probe should possess a G at either terminus.
32. All probe annealing sites should be free of SNPs.
33. The resulting amplicon should have a GC content of 40–60% and is ideally 150–200 bp in length [41]; however, we have successfully used amplicons up to approximately 300 bp.

References

1. Acharya JP, Lopez Y, Gouveia BT et al (2020) Breeding alfalfa (*Medicago sativa* L.) adapted to subtropical agroecosystems. Agronomy 10: 742
2. Singer SD, Weselake RJ, Acharya S (2018) Molecular enhancement of alfalfa: improving quality traits for superior livestock performance and reduced environmental impact. Crop Sci 58:55–71
3. Qin F, Liu D, Sun B et al (2016) Identification of alfalfa leaf diseases using image recognition technology. PLoS One 11:e0168274
4. Nutter FW Jr, Guan J, Gotlieb AR et al (2002) Quantifying alfalfa yield losses caused by foliar diseases in Iowa, Ohio, Wisconsin, and Vermont. Plant Dis 86:269–277
5. Undersander D, Cosgrove D, Cullen E et al (2011) Alfalfa management guide. American Society of Agronomy, Inc., Crop Science Society of America, Inc., and Soil Science Society of America, Inc, Madison
6. Xiao K, Kinkel LL, Samac DA (2002) Biological control of Phytophthora root rots on alfalfa and soybean with Streptomyces. Biol Control 23:285–295
7. Annicchiarico P, Barrett B, Brummer EC (2015) Achievements and challenges in improving temperate perennial forage legumes. Crit Rev Plant Sci 34:1–3
8. Lamb JF, Sheaffer CC, Rhodes LH et al (2006) Five decades of alfalfa cultivar improvement: impact on forage yield, persistence, and nutritive value. Crop Sci 46:902–909
9. Li X, Brummer EC (2012) Applied genetics and genomics in alfalfa breeding. Agronomy 2:40–61
10. Masoud SA, Zhu Q, Lamb C et al (1996) Constitutive expression of an inducible β-1, 3-glucanase in alfalfa reduces disease severity caused by the oomycete pathogen *Phytophthora megasperma* f. sp. medicaginis, but does not reduce disease severity of chitin-containing fungi. Transgenic Res 5:313–323
11. He XZ, Dixon RA (2000) Genetic manipulation of isoflavone 7-O-methyltransferase enhances biosynthesis of 4′-O-methylated isoflavonoid phytoalexins and disease resistance in alfalfa. Plant Cell 12:1689–1702
12. Yang S, Gao M, Xu C et al (2008) Alfalfa benefits from *Medicago truncatula*: the *RCT1* gene from *M. truncatula* confers broad-spectrum resistance to anthracnose in alfalfa. Proc Natl Acad Sci U S A 105:12164–12169
13. Gallego-Giraldo L, Jikumaru Y, Kamiya Y et al (2011) Selective lignin downregulation leads to constitutive defense response expression in alfalfa (*Medicago sativa* L.). New Phytol 190: 627–639
14. Stefanova G, Slavov S, Gecheff K et al (2013) Expression of recombinant human lactoferrin in transgenic alfalfa plants. Biol Plant 57:457–464
15. Tyagi S, Kumar R, Kumar V et al (2021) Engineering disease resistant plants through CRISPR-Cas9 technology. GM Crops Food 12:125–144
16. Peng A, Chen S, Lei T et al (2017) Engineering canker-resistant plants through CRISPR/Cas9-targeted editing of the susceptibility gene *CsLOB1* promoter in citrus. Plant Biotechnol J 15:1509–1519
17. Subedi U, Ozga JA, Chen G (2020) CRISPR/Cas-mediated genome editing for the improvement of oilseed crop productivity. Crit Rev Plant Sci 39:195–221
18. Mohanta TK, Bashir T, Hashem A et al (2017) Genome editing tools in plants Genes 8:399
19. Wolabu TW, Cong L, Park JJ et al (2020) Development of a highly efficient multiplex genome editing system in outcrossing tetraploid alfalfa (*Medicago sativa*). Front Plant Sci 11:1063
20. Chen H, Zeng Y, Yang Y et al (2020) Allele-aware chromosome-level genome assembly and efficient transgene-free genome editing for the autotetraploid cultivated alfalfa. Nat Commun 11:1–11
21. Zhou W, Liang G, Molloy P et al (2020) DNA methylation enables transposable element-driven genome expansion. Proc Natl Acad Sci U S A 117:19359–19366
22. Papikian A, Liu W, Gallego-Bartolomé J et al (2019) Site-specific manipulation of Arabidopsis loci using CRISPR-Cas9 SunTag systems. Nat Commun 10:1–1
23. Li J, Zhang X, Sun Y et al (2018) Efficient allelic replacement in rice by gene editing: a case study of the *NRT1.1B* gene. J Integ Plant Biol 60:536–540
24. Li Y, Zhu J, Wu H et al (2020) Precise base editing of non-allelic acetolactate synthase genes confers sulfonylurea herbicide resistance in maize. Crop J 8:449–456
25. Lin Q, Zong Y, Xue C et al (2020) Prime genome editing in rice and wheat. Nat Biotechnol 38:582–585

26. Singer SD, Laurie JD, Bilichak A et al (2021) Genetic variation and unintended risk in the context of old and new breeding techniques. Crit Rev Plant Sci 40:68–108
27. Jinek M, Chylinski K, Fonfara I et al (2012) A programmable dual-RNA-guided DNA endonuclease in adaptive bacterial immunity. Science 337:816–821
28. Meng X, Hu X, Liu Q et al (2018) Robust genome editing of CRISPR-Cas9 at NAG PAMs in rice. Sci China Life Sci 61:122–125
29. Pyott DE, Sheehan E, Molnar A (2016) Engineering of CRISPR/Cas9-mediated potyvirus resistance in transgene-free Arabidopsis plants. Mol Plant Pathol 17:1276–1288
30. Zhang S, Wang L, Zhao R et al (2018) Knockout of *SlMAPK3* reduced disease resistance to *Botrytis cinerea* in tomato plants. J Agric Food Chem 66:8949–8956
31. Kim YA, Moon H, Park CJ (2019) CRISPR/Cas9-targeted mutagenesis of *Os8N3* in rice to confer resistance to *Xanthomonas oryzae* pv. *Oryzae*. Rice 12:1–13
32. Tripathi L, Ntui VO, Tripathi JN (2020) CRISPR/Cas9-based genome editing of banana for disease resistance. Curr Opin Plant Biol 56:118–126
33. Li C, Li W, Zhou Z et al (2020) A new rice breeding method: CRISPR/Cas9 system editing of the *Xa13* promoter to cultivate transgene-free bacterial blight-resistant rice. Plant Biotechnol J 18:313–315
34. Zhang P, Du H, Wang J et al (2020) Multiplex CRISPR/Cas9-mediated metabolic engineering increases soya bean isoflavone content and resistance to soya bean mosaic virus. Plant Biotechnol J 18:1384–1395
35. Subedi U, Jayawardhane KN, Pan X et al (2020) The potential of genome editing for improving seed oil content and fatty acid composition in oilseed crops. Lipids 55:495–512
36. Xing HL, Dong L, Wang ZP et al (2014) A CRISPR/Cas9 toolkit for multiplex genome editing in plants. BMC Plant Biol 14:1–12
37. Tian L, Wang H, Wu K et al (2002) Efficient recovery of transgenic plants through organogenesis and embryogenesis using a cryptic promoter to drive marker gene expression. Plant Cell Rep 20:1181–1187
38. Badhan A, Jin L, Wang Y et al (2014) Expression of a fungal ferulic acid esterase in alfalfa modifies cell wall digestibility. Biotechnol Biofuels 7:1–5
39. Bae S, Park J, Kim JS (2014) Cas-OFFinder: a fast and versatile algorithm that searches for potential off-target sites of Cas9 RNA-guided endonucleases. Bioinformatics 30:1473–1475
40. Gao R, Feyissa BA, Croft M et al (2018) Gene editing by CRISPR/Cas9 in the obligatory outcrossing *Medicago sativa*. Planta 247: 1043–1050
41. Mock U, Hauber I, Fehse B (2016) Digital PCR to assess gene-editing frequencies (GEF-dPCR) mediated by designer nucleases. Nat Protoc 11:598
42. Doench JG, Hartenian E, Graham DB et al (2014) Rational design of highly active sgRNAs for CRISPR-Cas9–mediated gene inactivation. Nat Biotechnol 32:1262–1267
43. Graf R, Li X, Rajewsky K (2019) sgRNA sequence motifs blocking efficient CRISPR/Cas9-mediated gene editing. Cell Rep 26: 1098–1103
44. Gerashchenkov GA, Rozhnova NA, Kuluev BR et al (2020) Design of guide RNA for CRISPR/Cas plant genome editing. Mol Biol 54:24–42

Index

Nora A. Foroud and Jonathan A. D. Neilson (eds.), *Plant-Pathogen Interactions*, Methods in Molecular Biology, vol. 2659,
https://doi.org/10.1007/978-1-0716-3159-1,

N

O

P

Q

R

S

T

V

W

9781071631584